FIREFLY

101 OBJECTS
TO SEE IN THE NIGHT SKY

ROBIN SCAGELL

FIREFLY BOOKS

Robin Scagell is a long-serving Vice President of Britain's Society for Popular Astronomy (www.popastro. com). A lifelong stargazer, he has worked as an observer and photographer, and as a journalist has edited a wide range of popular-interest magazines. Robin is the author of several popular astronomy books, and has contributed to many other publications. He has been awarded the Sir Arthur Clarke Award for Space Reporting in recognition of his many appearances on TV and radio talking about astronomy and space.

A FIREFLY BOOK

Published by Firefly Books Ltd. 2014

Copyright © 2014 Robin Scagell

First printing

Publisher Cataloging-in-Publication Data (U.S.)

A CIP record for this title is available from the Library of Congress

Library and Archives Canada Cataloguing in Publication

A CIP record for this title is available from Library and Archives Canada

Robin Scagell has asserted his moral rights under the Copyright, Designs and Patents Act, 1988, to be identified as the author of this work.

Published in the United States by
Firefly Books (U.S.) Inc.
P. O. Box 1338, Ellicott Station
Buffalo, New York 14205

Published in Canada by
Firefly Books Ltd.
50 Staples Ave, Unit 1
Richmond Hill, Ontario L4B 0A7

Published in Great Britain by Philip's,
Endeavour House, 189 Shaftesbury Avenue,
London WC2H 8JY
An Hachette UK Company

Printed in China

Acknowledgments
All photographs supplied by Galaxy Picture Library.
All illustrations/photographs by Robin Scagell except as listed below:
Paul Andrew 90–91; Richard Bizley/Robin Scagell 107b, 119b, 125b; Adam Block/NOAO 131b; Juan Carlos Casado 79; Adrian Catterall 188; Celestron International 22tl, 22cr; Jamie Cooper 31t, 51, 68, 81, 109; Heather Couper/Nigel Henbest 80; A Fruchter/STScI 4l, 121; Eddie Guscott 141b,154t,154b,212,213t; Nick Hart 18l; David Hepwood 111; Nick Hewitt 195t; Mike Hezzlewood 197; James Jefferson 70; JPL 55, 62; Thierry Legault 16tl, 33b, 41, 45, 47, 200t; Margarita Karovska (Harvard-Smithsonian CfA)/NASA 191r; Martin Lewis 18r; Meade 22c; Don P. Mitchell 54r; NASA 16tr, 29, 37b, 85b; NASA/ESA/M. Buie (SWRI) 75; NASA/ESA/Hubble 105, 159; NASA/JHUPL/Carnegie Institution 53; NASA/JPL-Caltech 217; NASA/JPL-Caltech/ASI/Cornell 69b; NASA/JPL-Caltech/Malin Space Science Systems 57; NASA/JPL-Caltech/Space Science Institute 65r, 67b, 69t; NASA/JPL/University of Arizona 3, 60; NASA/Erich Karkoschka, University of Arizona 64, 71t, 73; NASA/NOAO/ M Meixner (STScI), & T A Rector (NRAO) 69t; NOAO/AURA/NSF 5l, 132r, 147, 171, 173, 183; Optical Vision Ltd 22tc, 22tr, 22cl, 24; Damian Peach 58, 65c, 67t, 125t, 157; Jeremy Perez 139t; Philip Perkins 174; Chris Picking 149; Sally Scagell 7; Peter Shah 4r, 142, 163, 198c, 198b, 207; Ian Sharp 39; N A Sharp, REU program/NOAO/AURA/NSF 175; Dave Tyler 49, 66; Ralf Vandebergh 85t; Doug Williams, REU Program/NOAO/AURA/NSF 5r, 185b.

Mapping (© Philip's)
Moon maps by John Murray.
Star maps (pp218–223) by Wil Tirion.

INTRODUCTION

This book is really for young astronomers who want to find something to look at in the night sky every time they go out to observe. But one or two older astronomers might find it useful as well. Why 101? Well, it's a good number, and they should keep you going for ages.

There is a point score for each object you find. There are no prizes, so there's no advantage in cheating here. It's up to you to keep tally and gain as many points as you can. It helps you to check how far you are getting through the objects, and is a reminder of what you've seen.

As you record your points, make a note of the date when you got them. It's also a good idea to keep a separate notebook, or keep a list on your computer, of what you observed and when. You could make drawings of some of the objects as well. This is a great way of remembering what you've seen, better than taking photographs really. The fainter objects are quite tricky to photograph, whereas a pencil and paper are all you need to make a quick sketch.

Now I know the title says "101 Objects to See in the Night Sky" and yet we have the Sun in there, which of course you can't see at night. And if you count all the objects, there are far more than 101 of them. But you can't really have a book about astronomical things to view and not talk about the Sun, and while you are finding 101 objects you can see others at the same time.

All the objects can be seen from North America. Many can be seen with the naked eye. This is a term that sometimes causes hilarity among young astronomers, and it just means "by eye alone, not using any binoculars or telescope." It's a term that astronomers use all the time, though some books prefer to say "unaided eye" because it doesn't sound so rude. But in this book, we tell it like it is. Actually, you'd be mad not to wear a lot of clothes – it gets very cold out there and you can concentrate much better when you're comfortable. So let's not have any complaints from parents that you caught a cold while using this book.

Most of the objects look better, or can only be seen, if you have binoculars or a telescope. These needn't cost very much, and you don't need anything grand to see any of them. Some are easier to see the farther south you live, and some are better the farther north you are. Some objects are really only visible if you are out in the country, so you may have to choose your moment. There are icons at the top of each object page to tell you what you need in order to observe each one.

This isn't a book about general astronomy, so to find out more about stars, galaxies and so on you may want to do some more background reading, or search online.

Have fun observing, and if you see everything in the book you'll be well on the way to being an expert observer.

Objects to look for in autumn   179

Objects to look for in summer 143

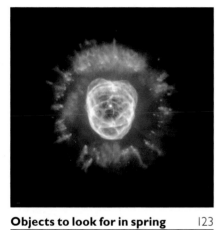

Objects to look for in spring 123

CONTENTS

THINGS YOU NEED TO KNOW

There is quite a bit you need to know first to be able to find objects in the sky. Obviously you need to know your way around the sky, and you need to know how to use binoculars and a telescope. You might have a computer-controlled Go To telescope, in which case you might think that finding objects is easy-peasy. But even then, you might need to know just what you are looking for, as not everything is in-your-face and obvious. So here are a few "things you need to know" so that you can understand how to find all 101 objects.

In each article, by the way, objects that have separate articles about them are referred to in **bold type**.

Constellations

These are names given to patterns of bright stars, and most of them got those names long ago. All the best-known ones date back to the Ancient Greeks (who crop up a lot in this book), and by Ancient I mean around 2,000 or more years ago. Some constellations date back well before that.

People often expect constellations to be like those join-the-dots pictures where it's obvious what they are from the arrangement of dots. Some of them look nothing like what they are meant to be. But the stars don't obligingly line up to make the shapes that we might want them to, so if people complain to you that Perseus doesn't look like any sort of hero, just point out that it would be odd if it did!

Many of them have tales linked with them, and you can imagine that before the days of TV and so on people would sit around a fire under the stars and would tell legends using the patterns as a sort of storyboard. Some of these legends – the ones I can repeat to a young audience, at any rate – are given in the articles.

The lines joining the stars that you see on maps are put there just to help you to spot the patterns. I prefer fewer lines rather than more. Sometimes they are used to join up all the bright stars in a constellation, but that can make the maps look like a spider's web.

Star names

Some stars have names, such as Polaris and Aldebaran. Mostly, these were given to them centuries ago and apply to bright stars only. Names are good, but there

▲ A photo of the stars of Virgo with the legendary figure superimposed in blue and the modern lines in yellow. The figure doesn't really fit the stars, and every artist draws their own version.

comes a point where it is hard to remember them all, let alone think of new ones for every tiny star there is.

So most stars are given other descriptions. The brighter stars are given Greek letters according to the constellation they are in. So the brightest is usually Alpha, the next brightest Beta, and so on. You don't have to know all the Greek letters, but here is a list together with their English versions. This method of naming stars is called the Bayer system, after the man who devised it centuries ago.

α	alpha	ν	nu
β	beta	ξ	xi
γ	gamma	o	omicron
δ	delta	π	pi
ε	epsilon	ρ	rho
ζ	zeta	σ	sigma
η	eta	τ	tau
θ or ϑ	theta	υ	upsilon
ι	iota	φ or φ	phi
κ	kappa	χ	chi
λ	lambda	ψ	psi
μ	mu	ω	omega

There's one catch with this system. The name of the constellation has to be included,

but it is used in the Latin possessive tense. So Delta of Gemini becomes Delta Geminorum. Much of the time this is shortened to Delta Gem, but you get the idea.

The bright stars all have Bayer letters as well as names. So Aldebaran in Taurus is also Alpha Tauri. And fainter stars in each constellation get numbers, such as 61 Cygni. After that there are all sorts of catalogs, and most stars don't have any names at all.

Incidentally, don't fall for the adverts that tell you that you can buy a name for a star. Not one name on any of those lists has any official status as they are not recognized by the International Astronomical Union. So save your money.

The difference between a star and a planet

That's easy. Stars are basically very distant suns, and have their own light, just like our own Sun does. Planets are much closer, orbit our own Sun, and don't have any light of their own so they are only seen by the light of the Sun. The Sun's planets are known as the Solar System, and they all go around the Sun in the same direction along paths close to the line of the ecliptic, which is shown on the maps on pages 218–223. In order outward, they are Mercury, Venus, Earth, Mars, Jupiter, Saturn, Uranus and Neptune. Other stars have their own planets but we can't see them directly because they are so faint and distant.

In the sky, stars stay in the same position all the time compared with one another, so the constellations are fixed and can be shown on maps. But the planets, being closer to us, don't stay fixed in the sky and move against the background of the constellations, so they can't be shown on printed maps

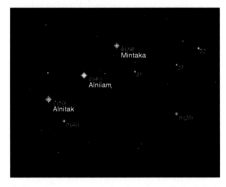

▲ The three stars in Orion known as Orion's Belt each have names, but they also have Greek Bayer letters and also numbers. The fainter stars just have numbers.

though computer maps can show them for any date and time. The article on each planet tells you where to find it in each year.

The stars are so distant (most are millions of times farther away than the planets) that they appear only as points of light in the sky, even though they are much larger than planets.

While we're at it, a moon is a smaller body that orbits a planet. Earth has just one, which we call the Moon, with a capital "M," but some planets have dozens of moons, with a small "m."

Where to observe from

The best place to start observing is where you live. Even in the very worst of light pollution you can always see something, even if it's only the brightest stars and the Sun, Moon and planets. So if this applies to you, get to know the sky as well as you can from home before you start trying to find a better site.

If you are using a telescope you ought to be outside, because the quality of the view is better if the telescope is at the same temperature as the surroundings. Air turbulence can play havoc with images, and if the telescope is warm on a cold night you can find that the images look as if you are seeing them through a fast-flowing stream. So let the telescope cool down, but don't leave the covers off the lenses or mirrors because they can get dewed up rather quickly. On some nights the turbulence – what astronomers call *seeing* – is bad anyway, because the upper atmosphere is turbulent, and there's nothing you can do about that.

But sometimes you can be better off indoors, observing through an open window, if it means that you can keep warm and maybe don't have a streetlight shining in your eyes.

▲ Twice every century, Venus crosses in front of the Sun. At these times, you get an idea of the difference in size between a planet and the Sun – though Venus was only about a quarter of the distance of the Sun from Earth.

Lights are a problem for many people. Ideally the only light you see should be from the sky. And choosing the right night to observe can make a big difference. Some nights are darker than others, and if you are aiming to see something faint, you need to wait for the darkest moonless night. Dust and water vapor in the atmosphere can spoil what promises to be a clear night. Those days when the sky is really deep blue can mean a good dark night ahead, even if there are some clouds as well. But a pale-blue milky sky by day is often only good for observing the Moon, planets and the brighter star clusters.

If you've spent the evening watching the TV or working on the computer and then look out to see if it's clear, don't expect to be able to see much straight away. Your eyes take a long time to get adapted to the darkness. Allow a few minutes in the dark to tell whether it's really clear, and at least 20 minutes if you want to see any deep-sky objects. To see your way around, use a red-covered torch or a rear cycle lamp rather than a white one, as red light doesn't affect your night vision as much as white light.

How the sky moves

You know from everyday life that the Sun and Moon rise and set. Read at the article on the Sun on page 92 to find out why most people are wrong about the Sun's rising and setting habits. But what's less obvious is that the stars rise and set in just the same way. The whole sky is constantly on the move, but of course it's really the Earth that does the turning, not the sky. If you could see it speeded up, you'd notice that everything rotates around a point in the sky in the north, not overhead. This is called the North Celestial Pole.

Just remember the following about the sky's daily motion:

- Looking north, the sky moves counterclockwise around the North Celestial Pole.
- Looking south, things move from left to right.
- Looking east, everything rises at an angle to the horizon.
- Looking west, everything sets at an angle to the horizon.

▲▶ These time exposures of the stars show how the sky moves looking in different directions. **1** Looking north, the stars turn counterclockwise around a point in the sky (which happens to be close to the star Polaris). **2** Looking east, stars (in this case Orion) rise at an angle to the horizon. **3** When Orion is in the south, its stars move horizontally, though in the far south they just rise and set in short arcs. **4** Looking west, stars set at the same angle that they rise.

▲ This fisheye view of the summer sky shows the large Summer Triangle of Vega, Deneb and Altair on the left (in the east) and the bright star Arcturus in the west. Compare it with the summer sky map on page 143.

But the Earth goes around the Sun as well, and this changes the view that we get each night, bit by bit. So the constellations on view change every few months, though those in the northern part of the sky are always the same but just different ways up. For each constellation you need to know when it will be visible, which is usually given in the "Where to look" section.

There are star maps on pages 218–223, showing the months when the various constellations are visible.

How to learn the sky

Once you've worked out how the sky moves, you can start to learn the stars and constellations. Do this in the same way that you learn your way around any new place – begin by finding a few landmarks, or in this case "skymarks," then fill in the gaps from there.

These skymarks are usually the bright stars or obvious patterns of stars. It's best to start learning them from your home site. Begin by getting your bearings – working out where north, south, east and west are. If you don't already know this, use a map, or notice where the Sun is at midday (in winter) or 1 pm (in summer), which is more or less due south.

To start with, scan the northern part of the sky for the Big Dipper or Cassiopeia, which are always up there somewhere – check out the diagrams here if you can't spot them straight away. Between the two of them is the Pole Star, which is almost exactly due north, so it's very handy. Now you know where north is, find where south, east and west are. It's a good idea to work out which bit of your horizon lies in each of those directions from your favorite observing spot.

Not everyone can see the whole sky from their home site, so you may have to find another spot to observe from to be able to see each bit. South is the most useful direction to observe, because the constellations and planets are at their highest

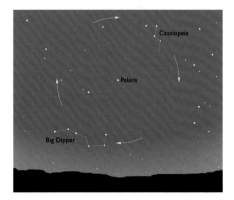

▲ Look north and you will see the Big Dipper and Cassiopeia, with the Pole Star, Polaris, between them. But they change positions according to the time of year.

when they are due south, and they parade past you from east to west (left to right).

Now you can start to find the stars and learn the sky. Use a computer program or phone app, a planisphere or the maps from pages 218 to 223. Look for the brightest stars and patterns. For each time of year there are particular skymarks. From these, move outward and pick out the other constellations. Bear in mind that there may be bright planets in the sky as well, which aren't on the printed maps but could change the view.

For example, in winter look for the constellation of Orion in the south and the bright star Capella overhead. In spring, it's

▲ Adjusting binoculars. **1** Get the best spacing to suit your eyes. **2** Focus using the left eye only. **3** Now focus using the right focuser only.

Leo in the south, with Arcturus the bright star farther east. Bluish-white Vega is overhead in summer, and the stars Deneb and Altair make a big triangle with it. By the fall, Vega has moved over to the west and the Square of Pegasus is the thing to look for, though it isn't very bright.

Whenever you start to observe, first check with the sky guides on pages 99, 123, 143 and 179 to get some ideas about what to look for. But the Sun, Moon and planets, and other Solar System bodies, can be visible at any time of year, so look for them separately.

Tips on using binoculars

Binoculars are easy enough to use – or are they? You'd be surprised how many people don't know how to use them and adjust them. There are two adjustments, in addition to the main focuser. One is getting the distance between the two eyepieces suited to your eyes, often by twisting the two halves around the central pivot. Make sure that each eye is looking exactly through the middle of each eyepiece. If there is a scale (which measures the separation in millimeters) around the pivot, make a note of your personal setting and then you can adjust them quickly each time you use them.

The other adjustment is on the right-hand eyepiece. But don't fiddle with this until you have got the left half of the binoculars in focus first. Choose some distant object and focus using the left half only, using the main binocular focuser (usually a wheel in the middle), then adjust the focuser on the right-hand eyepiece to bring the right half into focus. This corrects for any differences in strength between people's eyes, and may also have a scale so you can adjust it to your setting.

If you wear glasses you can probably use the binoculars without them, as the adjustments should be able to make the same corrections as your glasses. Only if you need the glasses for astigmatism will you need to keep them on, and the same applies to looking through a telescope. Often there is a rubber eyepiece shield which you can fold back so that you can press your glasses right up to the eyepiece.

▲ Choose a distinctive object when adjusting your finder to match the view in the main telescope.

Tips on using telescopes

There are thousands of telescopes that stay in their boxes on clear nights because their owners don't know how to use them and say that they can't see anything through them. But follow these two rules and you should be able to get almost any telescope working. If it works by day, it will work by night as well. If you don't already have a telescope and are thinking of getting one, take a look at our buyer's guide on page 21.

Rule number one is to make sure that the finder is lined up with the main telescope, even if you have a Go To telescope. The finder is the small device or telescope attached to the main tube, and it's there because the main telescope magnifies so much that you can easily miss the object you're looking for.

Align it by first finding an object with the main telescope, then adjusting the finder to show the same object. Do this by day, because it's easier to find objects then, and they generally stay still whereas objects in the sky are always on the move. Once you've got the finder aligned with the main scope, you can use it to point the main telescope accurately. Go To telescopes require you to begin by finding bright stars, so that they can then find other objects. But if your finder is not accurately aligned,

you won't see the same things through the main telescope.

Some telescopes have red-dot finders, in which you see the reflection of a red dot on the sky, while others are miniature telescopes. Either way, they all have adjusters, so make sure you can adjust them properly. The telescope variety usually give an upside-down image, which is often a drawback when comparing what you see with a star map, so you need some practice when using them to find faint objects.

Rule number two is that you should always start observing with your lowest magnification, which is the eyepiece with the highest number on it. That will give you the widest and brightest field of view. You can increase the magnification by using different eyepieces once you have located what you are looking for, but finding it first can be the challenge.

And finally (though I could think of many more rules if requested), try to get the most out of what you can see with low magnifications before going to higher ones. Increasing the magnification not only makes things bigger, but it also makes many things fainter because the same amount of light is spread over a larger area. You only see a smaller part of that area as well. If the object you were looking at was not dead center before you changed the eyepiece, and particularly if you don't have a motor drive, it may not even be in the field of view at all. And to make things worse, the new eyepiece may need to be refocused, so you don't see a thing if there are only faint stars in the field of view! It's not so bad on the Moon, where you can usually see something, but on some faint nebula or other increasing the magnification can be a real trial.

Telescope mountings often cause trouble, particularly on the cheaper scopes. Getting the telescope to stay put where you want to look can be tricky, and trying to follow the object through the sky is even worse because the controls are not smooth enough. One tip that might just help is to provide a bit of tension on the telescope using a bungee cord over the tube over the pivot point, secured to the tripod. This could prevent the telescope from wobbling on its bearings if they are worn, but there are no guarantees.

▲ This elderly and battered 60 mm refractor is difficult to use because the bearings have worn. But a bungee over the tube helps to keep the image steady.

Viewing the wrong way round

Most astronomical telescopes give an upside-down, or inverted, image, because this uses the least number of extra lenses. So you need your wits about you when comparing what you see through a telescope with a sky map or Moon map. To make matters more complicated, many telescopes use a star diagonal, which gives a mirror image as well!

These star diagonals are used to make the viewing angle easier, but sometimes they are essential to help get the telescope to focus. If you can't focus the telescope even by day because the focusing mechanism doesn't travel out far enough,

◄ What can happen when you increase the magnification.
1 A view of the Crab Nebula with a magnification of about 35 times.
2 Increasing the magnification to 70 times throws the stars out of focus and if you're lucky all you see is a few faint circles of light.
3 Focusing the telescope makes the stars sharp again – but the nebula itself is just outside the field of view.
4 This is what you wanted in the first place – a magnified view of the nebula. Actually, the photo shows more detail that you could see by eye anyway.

you probably need to use the star diagonal between the telescope and the eyepiece.

East and west

This is another thing that can cause complications. When you look up in the sky, facing south, east is to your left and west is to your right, and these directions are often used when explaining where to find one object compared with another. So Taurus is to the northwest of Orion, for example. But this means that when you look at a sky map, bear in mind that east and west are the opposite way round from what you would expect, with west to the left and east to the right.

At least north and south are always the same. Or are they? When objects are rising or setting, they do so at an angle to the horizon. When Orion is rising, Taurus is above it. So while it is still to Orion's northwest, as usual, it is not to its upper right as you might imagine. And when Orion is setting, Taurus is to its right.

Just when you have got your head around that, there is another complication. When you look at the Moon, you have to think of directions not as you would in the sky, which is based on Earth's east and west, but as if you were an astronaut standing on the Moon. So when talking about directions on the Moon, they are the same as on an Earth map, with west on the left.

Angles in the sky

Describing positions in the sky involves using angles. Normally, when we want to say where one thing is compared with another, we say things like "He lives five houses down the road" or "About 10 cm away from that one." But these depend on houses all being roughly the same, and knowing what distance the objects are from us. Stars are faint and bright, and we have no idea when we look how far apart they are. So we use angles, which just refer to the distance around a circle.

A full circle is 360°, and the distance from the horizon to the overhead point, called the *zenith*, is 90°. Halfway up the sky is 45°. One simple way to measure an angle is to hold your outstretched hand at arm's length. This covers an angle of around 18°. Small people have smaller

◀▲ When Orion rises in the east in late autumn, the Belt stars are at a steep angle to the horizon and point to Aldebaran and the Hyades in Taurus, almost directly above Orion. The bright object is Jupiter. But in late spring Orion is setting in the west and the Belt stars are almost horizontal. Taurus is now to the right of Orion.

15

▲▶ The Hadley Rille, as seen from Earth and from the Moon's surface. From Earth, though our west is to the right, from an astronaut's point of view west is to the left of this view.

hands and shorter arms, so it works for most of the time! Your index finger covers just over 1°. Occasionally this sort of description is used in this book, so remember that it means "at arm's length."

Each degree is divided into 60 minutes, usually called arc minutes to distinguish them from time minutes, and each minute is divided into 60 arc seconds.

A few useful distances in angles are:

- Field of view of ordinary binoculars – about 5°–9°. In this book I use 5° as typical.
- Field of view of typical low-magnification eyepiece – about 1°.
- Sun or Moon diameter – ½° or 30 arc minutes.
- Diameter of Jupiter at its closest – 47 arc seconds.
- Diameter of Mars at its closest – 25 arc seconds.

Measuring star brightness

It was the Ancient Greeks who first started to measure the brightness of stars, and they simply called the brightest first magnitude and the faintest sixth magnitude. We still do the same today, more or less, but now we have put the scale on a scientific basis and have extended it to faint stars that the Ancient Greeks could never have seen. The scale also covers very bright objects, such as Venus and the Moon, which have negative magnitudes, and a few bright stars have magnitudes around zero. One big problem with the magnitude scale is that it runs the wrong way from other measurement scales, in which the more of something an object has, the bigger the number.

Here are a few typical magnitudes:

- Sun −26.8
- Moon −12.7
- Venus at its brightest −4.9
- Jupiter or Mars at brightest −2.9
- Sirius (brightest star) −1.5
- Vega 0.0
- Spica 1.0
- Polaris 2.0
- Zeta Tauri 3.0
- Typical faintest star visible from an urban area 3 to 4
- Alcor 4.0
- Phi Cassiopeiae 5.0
- Typical faintest star visible from a country area 6
- Faintest star visible with 10 × 50 binoculars 10 (though 9 is a practical limit in average skies)

- Faintest star visible with 80 mm telescope 12
- Faintest star visible with 130 mm telescope 13
- Faintest star visible with Hubble Space Telescope 31

If you are a unbearably bright spark at physics you may say: "This is all very well, but what are the units of magnitude? You can't have just a number. And what is the standard point of the scale, like temperature is measured against the freezing and boiling points of water?" Well, there are no actual units to the system but you could convert them into watts per square meter of light received at Earth's surface if you wanted to, though no one usually does. And nor is there a standard star, because no star can be relied upon not to vary even slightly. The scale is based on all the existing magnitudes that have been measured.

Visibility of deep-sky objects

What we call deep-sky objects include all the things such as star clusters, nebulae and galaxies beyond the Solar System that aren't actually stars. Star clusters and nebulae (gas clouds) are all part of an enormous collection of stars that we call the **Milky Way** Galaxy. Star clusters can be fairly small *open clusters*, usually with a few dozen or hundred members, or larger ones on the fringes of the Milky Way, called *globular clusters*. There are other galaxies as well, beyond the Milky Way, so distant that we can't see individual stars in them and they appear as just a misty blur.

Deep-sky objects are often the most interesting things to find, but they can be quite a challenge to spot. Although books often quote a brightness for them in terms

of their magnitude, as with stars, this isn't always the best guide to their visibility.

A large object might have the same total brightness as a small one, but because its light is spread out over a lot of sky, it appears dimmer. So in this book, rather than give magnitudes, I have devised a scale of visibility, where visibility A is easily seen from a built-up area, down to E, which is quite a challenge unless you have a good, dark sky. This assumes that you are using either binoculars or a small telescope (up to about 130 mm aperture).

Whatever instrument you use, even a large telescope, don't expect to see deep-sky objects looking exactly as they do

Mars	-2
Aldebaran	+1
Hamal	+2
Beta Trianguli	+3
Gamma Trianguli	+4
Phi Tauri	+5
SAO93918	+6

▲ The planet Mars, seen against the background of the winter stars of Taurus, Aries and Triangulum, providing a range of different magnitudes. See if you can spot all the stars – though Mars probably won't be there!

in photos. Your eyes work in a different way from cameras, which can build up light over many minutes, making objects look bright and bringing out their color. Most deep-sky objects appear very pale and colorless through a telescope. But it's still great fun finding them and seeing them for yourself.

If you don't see much when you look for a deep-sky object, bear in mind the tip on page 110 of using averted vision. Give things a fair chance when viewing, and if you can't see anything at all try on another occasion when the object is at its highest in the sky and it is just that bit clearer. These faint fuzzy objects are easily wiped out altogether by light pollution or haze, so wait until the night is really sparkly (which often means cold as well) before giving up on them.

For many people the hardest part comes once you've found the object. "Huh, is that it?", you might say, and move on to something else. But this is the point at which expert observers take a bit more time to really study what they can see, and even make a sketch. This might seem a bit old-fashioned, but it's actually quite hard to photograph these things and a sketch makes you concentrate on what you're seeing. Keeping a record of what the object looks like is really useful, because

you can compare the view with different instruments and from different locations. Many experienced observers keep observing books with records of what they saw right back to when they began.

Many of the photos in this book were taken by amateur astronomers, like the one shown below, often with fairly small telescopes. But they require expensive sensitive cameras and other equipment, so you won't be able to take similar shots on your phone camera!

Details of objects

You might think that everything in astronomy is pinned down exactly. Astronomers can tell you to the second when an eclipse will occur, while space scientists can send a probe to Pluto and say what day it will arrive there many years later. But other things are much less well known.

Distances to stars and galaxies, for example, can be terribly vague, because there's usually no direct way of measuring them. Astronomers just have to make estimates based on what they look like and by comparing them with similar objects

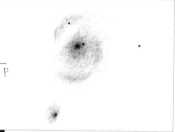

◀▲ The Whirlpool Galaxy in Canes Venatici is visible with small telescopes and even binoculars, but you don't get a view anything like that in the photos. The picture (left) was taken using a 250 mm telescope from light-polluted Newport, South Wales, while the drawing (above) was made using a 250 mm Dobsonian telescope from a dark-sky site.

whose distances are better known. So the details shown here may turn out to be a bit wrong, but they are there to give an idea of the distance rather than being a precise measurement.

Distances within the Solar System are usually given in kilometers, but beyond the Solar System they are so vast that another unit is needed: the *light year*. This is the distance that light travels in a year, and is equal to just under 10 trillion kilometers, or 10,000,000,000,000 km. No point in trying to imagine this distance, but at least the numbers get more manageable when given in light years!

Finding things in the sky

It's all very well being told that the stars of Pegasus are around in autumn, or that Saturn is in Ophiuchus, but how do you work out where and when to look? Even if you have a clever Go To telescope, or a smartphone app that shows you the sky at the press of a button, you could waste a lot of time trying to find Saturn in Ophiuchus in December when that part of the sky is just not visible then!

So this is where our star maps come in handy. Turn to pages 218 to 223 and you'll find simple maps of the evening sky. These show you the main constellations and stars visible, so you can work out when is the best time to observe. An alternative is to buy a *Firefly Planisphere*, a simple device which allows you to dial the date and time, and see a simple map of the sky for that moment. Many astronomers still use these even though they may have more advanced computer maps at their fingertips because they are so quick and easy to use.

These days, it is easy enough to get detailed star maps on your computer. One popular free download is Stellarium,

available from www.stellarium.org. Once installed, this gives you a very realistic view of the sky. Before using it, choose your nearest location from the list provided, or enter your own latitude and longitude. Bear in mind that you need to use a keyboard to control the movements – keep pressing L to move forward in time at increasing speed or J to move backward, and K to stop the movement. Take care or you will speed too far forward or back, in which case press 8 to take you back to the sky Now. Have fun exploring the various possibilities.

Many tablets and phones have apps which will show the sky visible at that moment or at any time, and if the device has the right bits inside the display moves as you hold it up to the sky. These are a great help in finding your way around the sky, but sometimes you have to shake them to make them work properly!

You might see positions quoted in Right Ascension and Declination. These are the sky equivalents of longitude and latitude, but the RA scale runs in hour and minutes, which relate to the turning of the sky, and run from west to east.

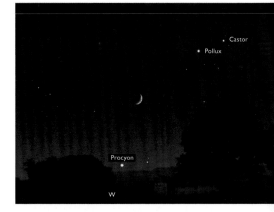

▲ Stellarium shows the sky from any place and at any time, with a realistic horizon and sky appearance. As you zoom in, more stars and deep-sky objects appear.

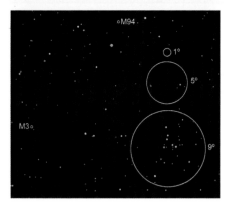

◀ To find the bright globular cluster M3 (shown in the small blue circle) you need to start at the Coma Star Cluster, which is inside the large white circle. This circle shows the 9° field of view of a typical 6 × 30 finder, the smaller one is a 5° field of view such as you'd get from binoculars, and the small circle is a 1° field of view typical of a telescope. Only the brighter stars are visible with the naked eye. Using either of the finders you should see enough stars to move from the Coma Star Cluster to M3 fairly easily in a few jumps, but with a telescope alone you would need about 20 separate jumps.

Star-hopping

Having worked out where you need to look, you then have to find the actual objects described. If you have a Go To telescope you can simply tell it to do the work, which is fair enough, but it does take a lot of the fun out of it. Those without Go To will have to find the objects for themselves. For those objects which you can't find using the naked eye alone, you'll need to become good at star-hopping using binoculars or your telescope.

The method of star-hopping involves finding a known star to start with, then moving bit by bit from there, finding other stars that may be visible only with binoculars until you reach your target. Many telescopes have finder telescopes with a field of view similar to that of binoculars, so you can use those instead. But smaller telescopes these days tend to have those red-dot finders, which some people love but I find rather useless, particularly in an area with light pollution, because they don't show any more stars than you can see with the naked eye.

With binoculars or a finder scope, you get an enlarged view of the sky and you can see more stars. They have a field of view from 9° for a six-magnification finder or binoculars to about 5° for a nine-magnification finder or binoculars. If your telescope just has a red-dot finder, use binoculars if you can first to find the part of the sky where your target lies, then try to aim the red dot at the same place. If you have no binoculars but just a telescope, you'll have to guess where it is compared with the nearby stars. If it's a long way from any stars that you can see, this is tricky and you'll have to use the telescope itself to go from star to star, which calls for a lot of patience.

About binoculars

Binoculars are very useful for astronomy. Even advanced observers usually have binoculars to hand for a quick view of the sky or just for general stargazing. The great thing about them is that they can be used

▲ A range of binoculars. Those at the back are 15 × 70 and 20 × 80, while at the front are the more convenient 8 × 42 and 10 × 50.

for daytime viewing as well, so you can get a lot of use out of them.

Most types of binoculars will help you see things in the sky, and you don't need to get a super-powered model. Binoculars are described using two numbers, such as 8 × 40. The 8 is the magnification and the 40 is the diameter of the main lenses in millimeters. The larger the lenses, the heavier the binoculars and the more difficult they are to hold for a long time, particularly when looking up at the sky. For most people, 8 × 40s are a very handy size. The larger ones, such as 10 × 50s, will show more deep-sky objects and fainter stars, but any larger than that and you are into the heavy stuff.

One good thing is that you don't need expensive all-weather or waterproof binoculars for astronomy – there's not much observing to be done when it's raining!

About telescopes

There are hundreds of different telescopes on sale, so how do you decide which to get? There are three main types: refractors, reflectors and catadioptrics. All usually give upside-down images, which doesn't matter much for astronomy but is a pain when you want to use them by day.

Refractors are the standard telescope that everyone knows, with a lens at the top end of the tube and an eyepiece at the bottom. They are easy to use and good value in the smaller sizes, but the cheaper ones tend to give colored fringes around bright objects, particularly when used at high magnification. The viewing angle can be awkward unless you use the star diagonal. They can be used by day with the addition of an extra lens or prism.

Reflectors have a mirror at the bottom of the tube instead of a lens at the top, and

you view the image looking sideways into the tube. This can be more comfortable. The mirror does not give colored fringes, but its coating can tarnish over time and is tricky to clean. Mirrors can be made larger than lenses, so the larger sizes of telescope are usually reflectors. The standard type of reflector is called a Newtonian telescope.

Catadioptric telescopes have a mirror at the bottom and a glass corrector plate at the top, and you view up through the tube as with a refractor. They are more compact than either of the other two types, but cost more.

Telescope mountings

The telescope mounting is just as important as the telescope. When you are observing at a magnification of 100, say, every little jiggle is magnified by 100 times, and the object you're looking at moves through the field of view amazingly quickly, so a good mount will keep the scope steady and allow you to follow objects easily using either manual *slow motions* or a motor drive. Given the choice, I'd have a motor drive any day (or night).

The cheapest mounts are usually rather wobbly, with sticky slow motions, and require lots of practice and patience. It's probably better to get a more firmly mounted scope. There are a few types of mounting:

Altazimuth mounts allow you to move the scope from side to side and up and down. This seems logical, but objects in the sky usually move diagonally, so you have to keep shifting the mount in both directions by either just tugging the telescope, or using slow motions or motors.

There is a type of altazimuth mount called a Dobsonian, which uses low-friction pads to make the scope move smoothly. This is the simplest way of mounting a

1 A typical 70 mm refracting telescope on an altazimuth mount. At the bottom end of the tube you can see the star diagonal that gives a more comfortable viewing angle rather than having to crouch underneath the telescope looking upward.

2 This 150 mm refracting telescope is on an equatorial mount.

3 A 130 mm Newtonian reflecting telescope on a simple Dobsonian mounting. You can see the small secondary mirror that reflects the light to one side where you can view it, but the open tube design means that stray light from your surroundings can spoil the view.

4 A 150 mm Newtonian reflector on an equatorial mount. In this case the tube is closed, so it's easier to use in a light-polluted area. With this telescope the tube is in cradles so you can move the eyepiece to view at a comfortable angle, but always looking sideways on to the object you're viewing.

5 A 90 mm Maksutov-Cassegrain catadioptric telescope on an altazimuth fork Go To mount. This is a very portable instrument, but the aperture is rather small. The handset for controlling the telescope is not shown. Like the refractor, the eyepiece is at the bottom end of the tube, but you usually need a star diagonal to give a comfortable viewing angle.

6 A 150 mm Schmidt-Cassegrain catadioptric telescope on a single-arm altazimuth Go To mount. The tube is short compared with a Newtonian reflector of the same aperture, so it's more portable.

reflector, and Dobsonian-mounted reflectors are often the cheapest way to get more telescope for your money.

The only altazimuth mounts with motors are computer-controlled ones, because they need to know where in the sky they are looking so they can run the motors at the correct rate. These have to be aligned on the sky before they will work properly, which takes a bit of time to do. And you need batteries!

Equatorial mounts can be aligned so as to follow the sky using just one movement. Again, they take a bit of time to set up properly, and many beginners find them thoroughly confusing. In my view they are only worth having if the mount has a motor as well, then they can be really useful.

Go To telescopes have an inbuilt computer with a knowledge of the sky and motor drives, and are now used on both altazimuth and equatorial mounts. They sound ideal for beginners, as they will find objects for you at the touch of a button, but all except the most expensive versions require you to do some setting up beforehand, and if you get it wrong they are hopeless. They can be very useful, but are no substitute for knowing the sky.

Telescope jargon

The ads are full of jargon which you may want to understand, so here's a quick guide.

Aperture is the diameter of the mirror or lens.

Focal length is the distance between the lens or mirror and the image it gives. It is roughly the length of the telescope, in the case of refractors, so a telescope with a focal length of 600 mm is actually about 600 mm long. Reflectors are a little shorter, and catadioptrics are shorter

still, about a quarter or a fifth of the focal length. The longer the focal length, the more magnification you get with a standard eyepiece. Eyepieces have their own focal lengths – more about this shortly.

Focal ratio measures the focal length compared with the aperture. The range is about f/4 to about f/14. An f/4 telescope is short for its focal length, and is more suited to deep-sky observing in dark locations, while an f/14 telescope (usually a Maksutov catadioptric) is ideal for planetary observing and for observing in more light-polluted areas. But you could still use either telescope in either location. A telescope of about f/6 is a general-purpose instrument.

Eyepieces are what you use to look at the image and change the magnification. Each eyepiece has a focal length, usually in the range of 40 mm down to 4 mm. A standard eyepiece is usually around 25 mm. The magnification given by an eyepiece depends on the focal length of the telescope you use it with, because magnification is given by the focal length of the telescope divided by the focal length of the eyepiece. So if your telescope has a focal length of (let's make it easy) 1,000 mm, and the eyepiece is 25 mm, the magnification is 40, sometimes written 40×, meaning 40 times. This is true whatever the telescope's aperture, though a larger aperture will give a brighter view than a smaller one of the same focal length.

So if your telescope had a focal length of 600 mm and you used the same 25 mm eyepiece, you'd get a magnification of – no, I'm not going to tell you, you work it out. The answer is on page 24. And while you're at it, work out what eyepiece you'd need on the telescope with 1,000 mm focal length to get a magnification of 100.

▲ A range of different eyepieces. Usually the longer ones, with the longer focal lengths, give the lowest magnification.

A **Barlow lens** is included in the equipment of many telescopes. This has the same barrel as an eyepiece, but you can't use it by itself to view through the telescope. Its purpose is to multiply the power of any eyepiece that you use it with, by putting it in between the telescope and the eyepiece itself. A 2× Barlow, for example will double the magnification of any eyepiece. So if you have two eyepieces and one Barlow, you have four different magnifications.

The Barlow lenses sold with the lower-priced telescopes are often not of good quality, and introduce false color into the view, so don't rely on them too much. Rather than buy a higher-power eyepiece, getting a better Barlow lens might be a wise move.

One tip is that if you use a star diagonal (or any other extension tube) between the Barlow and the eyepiece, you increase the magnification even more. So you could get three different magnifications from one eyepiece.

OTA means Optical Tube Assembly – the tube of the telescope alone, without a mount.

What do the symbols mean?

Many of the 101 objects can be seen with the naked eye with no trouble. For some, at least binoculars are needed, while for others you really need a telescope, though not a particularly large one. With the heading for each object there is a symbol to give you a guide to its visibility. Many can be seen with all three, showing that you can see more detail with binoculars or a telescope. Here's what they mean.

 Naked eye under average conditions.

 Ordinary binoculars needed.

A small telescope (no larger than 130 mm aperture) is needed.

▲ This 130 mm refractor is on an equatorial mount. The time exposure shows how the stars have trailed around the Pole Star, and the telescope axis is pointing at the same spot. This means that it will follow the stars using just one movement.

Answers to the magnification questions on page 23. A 25 mm eyepiece on a 600 mm focal length telescope gives a magnification of 24. To get a magnification of 100 from a 1,000 mm focal-length telescope you'd need a 10 mm eyepiece.

THE SOLAR SYSTEM

Over a third of the 101 objects in this book are in the Solar System. There is no particular time of year when you can see most of these, so you're going to have to put up with the fact that you may have to wait to see them.

This plan of the Solar System is not to scale – the orbits of Mercury out to Mars are very close together and the actual sizes of the planets are not to scale with the sizes of their orbits. The planets are always on the move so it just shows the order of the planets outward from the Sun, and a typical comet orbit, rather than their actual positions.

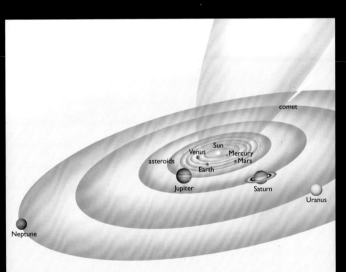

I

The Moon and its phases

The Moon is our only natural satellite and people take it for granted – but its changing phases govern the lives of astronomers!

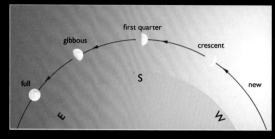

◀ ▼ How the Moon's phases appear from Earth. Reading from right to left, the Moon first appears as a thin crescent in the west after sunset, then after a few days is at first quarter. It grows in phase to gibbous, rising before sunset and often visible in the daytime sky. At full the Moon rises as the Sun is setting, but opposite it in the sky.

Where to find it

Sometimes the Moon is there, sometimes it isn't – so how do you know where to look for it? It also continually changes its appearance, or *phase*, which is caused by the varying angle that the Sun shines on it as the Moon orbits the Earth. Though the Moon goes round the Earth from west to east every 29.5 days, the Earth turns in the same direction much more quickly. If you look at it at roughly the same time day after day, you can see its changing position compared with the Sun. It goes through its complete range of phases in just less than about a month.

Fact File

Diameter	3,474 km
Period of orbit around Earth	27d 7h 43m
Synodic period (one full Moon to the next)	29d 12h 44m

Once you know the way its phases change, you can keep track of the Moon quite easily. When it is a crescent in the evening sky, it is always over toward the west, following the setting Sun. A few days later it is a half Moon, and is more or less due south or southwest an hour or so after sunset.

A week after this the Moon is full, and rises opposite the setting Sun at sunset. Another week passes and the Moon is at half phase again, but this time in the morning sky. Then it becomes a crescent again and rises in the eastern sky just before the Sun.

What you'll see

The Moon has no light of its own, and all we see is the sunlit portion. If you hold a tennis ball more or less in line with a lamp, only one edge of it will be illuminated, just like a crescent Moon. Hold it to one side and it is half illuminated, and when you hold it with the lamp behind you, it is fully illuminated, like full Moon.

The boundary between light and dark on the Moon is called the *terminator* –

▼ Two views of the same area of the Moon, with Mare Serenitatis at upper right, at first quarter (left) and full Moon. You can see much more detail in the terminator at first quarter, though lunar light and dark areas show up better at full Moon.

Point Score

●●●●●●●●●●	Date seen	Points
Seeing the waxing crescent Moon		1
Seeing the first quarter Moon		1
Seeing the gibbous waxing Moon		1
Seeing the full Moon		1
Seeing the gibbous waning Moon		1
Seeing the last quarter Moon		3
Seeing the waning crescent Moon		2
Seeing Earthshine		1
Score		

See also: **Mare Crisium; Mare Tranquillitatis; Mare Serenitatis; Mare Fecunditatis; Mare Imbrium; Oceanus Procellarum and Copernicus; Sinus Iridum; Aristarchus; Theophilus, Cyrillus and Catharina; Mare Nubium and the Straight Wall; Tycho and Clavius.**

nothing to do with those films where a robot goes around shooting people! Through a telescope or binoculars, details close to the terminator show up very clearly, because the Sun casts long shadows and shows up all the small lumps and bumps which you'd otherwise miss.

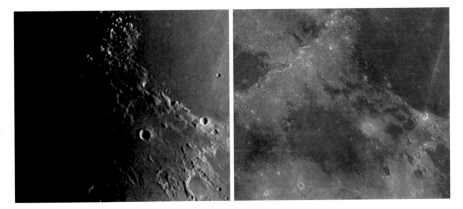

▲ A familiar sight – but the Moon is a world in its own right and as its phases vary new features come into view.

We call the crescent Moon a "New Moon," though strictly speaking this is when the Moon is in line with the Sun so we can't normally see it. After a few days the Moon is a crescent, and is still quite close to the Sun in the sky, so it sets soon after sunset. Spring is the best time to see the crescent Moon, because its path is high in the sky at that time of year. The half Moon is also called "first quarter" because it is now a quarter of the way around its orbit.

The evenings around first quarter are many people's favorite time to view the Moon, because so much detail is visible.

As the Moon gets more illuminated each evening, its illuminated portion gets fatter. This is known as the *gibbous* phase. It can be quite hard to tell when the Moon is actually full, because it looks almost complete for a few days on either side of the actual moment. Full Moon is often disliked by astronomers because its light drowns out most of the fainter stars and objects we want to see! And through a telescope all you see on the Moon itself are patches of light and dark, so details are quite hard to make out.

In summer, the full Moon is low in the evening sky, but in winter it is high – the opposite of the Sun's path through the sky at those times of the year.

After full Moon, the shadows start to return and it becomes gibbous again. But by now it is rising later and later each night, so it becomes more tricky to observe it at a reasonable time in the evening. You can just see a last quarter Moon before midnight in the fall, but for the rest of the year it is usually an early morning job, though early risers can see it around the time of sunrise. For most people this would be in winter, when the Sun rises as they are getting up.

The final crescent Moon in the morning sky before sunrise is sometimes called the "Old Moon." You might see this when you get up on winter mornings before sunrise.

Between new Moon and full, when the Moon's phase is growing, it is said to be waxing. After full Moon, its phase is shrinking, and it is waning.

Often when the Moon is a crescent you can see the whole of it faintly illuminated, even though the Sun isn't shining on it. The Earth itself is providing the illumination, and this is called *Earthshine*. If you were standing on the Moon at this time, you would see a full Earth hanging in the sky shining down on you.

About the Moon

We are so used to the Moon being in our sky that we don't really think of it as another world. Actually, most of the other moons of planets are very much smaller than the planet itself, whereas our Moon is about a third of the diameter of its parent planet. It was probably created when the Solar System was still forming and another body hit the Earth. The Moon formed out of the material in the splash.

Today it is a dead world, with virtually no atmosphere, and its once-molten core is now solid. It has been battered by collisions with asteroids since its formation 4.5 billion years ago. In the distant past, the biggest impacts released molten lava, which formed the darker areas we now call "seas."

One thing about the Moon is so familiar to us that we never question it. Apart from the phases, we always see the same markings. Yet the Moon goes round the Earth, so why don't we sometimes see different sides of it, and why doesn't it rotate? The answer is that it has *captured rotation*, which means that it rotates on its axis in exactly the same time that it takes to orbit Earth. Imagine someone walking counterclockwise in a circle around you, spinning counterclockwise as they do so. You would see every side of them. But if they slowed down their spin, eventually they would face you all the time, like the Moon does. Most of the large moons of the other planets do the same thing.

People often talk about "the dark side of the Moon" meaning the far side, but really the far side gets just as much sunlight as the near side. At new Moon, the far side is facing the Sun, for example. The actual far side can only be seen from spacecraft. It is covered with craters, like the near side, but has only two small dark areas, or seas.

The Moon is the only other world on which humans have set foot, during the six Apollo landings from 1969 to 1972.

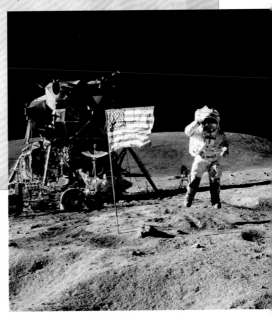

► Apollo 16 astronaut John Young jumps off the lunar surface while saluting. His heavy spacesuit would have made this difficult in Earth's gravity, but the Moon's gravity is only one-sixth that on Earth. In the background is the Lunar Module which took him and fellow astronaut Charles Duke to the Moon, and also the Lunar Rover which they used to explore their surroundings.

2

Mare Crisium

This lunar plain is one of the most easily recognized features on the Moon, and is a good starting point when learning your way around the lunar surface.

Where to look

Almost whenever you see the Moon in the evening sky, you can spot Mare Crisium, which is a dark oval area to the upper right of the Moon's disk. Even when the Moon is all but the thinnest crescent it is visible, as it is quite close to the edge, or limb, as seen from Earth. However, after full Moon it starts to move into shadow, and it can't be seen more than four days after full.

Because the Moon always keeps the same face pointing toward the Earth, Mare Crisium remains more or less in the same place as you look at it. Its name means "Sea of Crises."

What you'll see

Like many other features on the Moon, Mare Crisium was formed when something rather large hit the Moon billions of years ago when the Solar System itself was young.

The impact melted the rock and created a huge lake of liquid molten rock, which slowly solidified into the dark surface we see today.

Most people can see this dark oval without any optical aid, but the more magnification you use, the more you can see. With binoculars you can make out the outline more clearly, and also see a roughly diamond-shaped gray area to its left with a bright crater at one point of the triangle. This is the crater Proclus, 28 km across, which was formed like the other craters on the Moon when an asteroid hit the Moon's

Point Score

●●●●●●●●●●	Date seen	Points
Finding Mare Crisium		1
Finding Proclus		2
Finding Picard		3
Score		

Fact File

Name	Diameter
Mare Crisium	550 km
Proclus	28 km
Picard	23 km
Peirce	19 km
Swift	11 km

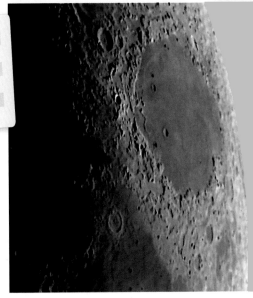

surface. The bright splash marks, or rays, from this impact form the two sides of the diamond, an area romantically known as Palus Somni, the "Marsh of Sleep." Because the rays only appear on one side of the crater, it's quite likely that the impact was quite a glancing blow, so the debris went mostly in one direction.

A telescope will reveal craters on the floor of the sea, the largest being Picard, Peirce and Swift. Picard is named after French 18th-century astronomer Jean Picard, not Jean-Luc Picard who was the fictional commander of the *Starship Enterprise*!

Although Mare Crisium appears oval because it is foreshortened as we see it from Earth, spacecraft photos show that it is actually elongated from east to west. It would be roughly circular but for the bay on the side closest to the limb, which extends it from being circular.

▲ Mare Crisium, as seen when the Moon is a thin crescent.

Mare Crisium and libration

Mare Crisium's location near the Moon's limb makes it a useful guide to what's known as libration – not to be confused with liberation! Although the Moon keeps the same face turned toward Earth at all times, there is a slight extra rocking movement, libration, which allows us to see a bit more of one edge or the other at different times. In all, you can see up to 59% of the Moon at one time or another. So at some times, Mare Crisium will appear to be closer to the limb than at others.

◀ The Moon at two different librations. Notice how the features shown by red dots are at different positions on the disk.

3

Mare Fecunditatis 👁️ 🔭

Positioned on the Moon's eastern side, this is one of the first lunar seas to be seen after new Moon.

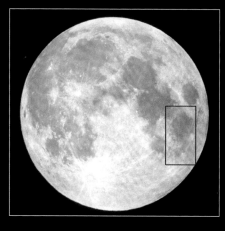

Where to look

Just a few days after new Moon, as soon as the Moon is no longer the thinnest crescent, you can start to see Mare Fecunditatis, the "Sea of Fertility." At this time, the Sun is just rising across its surface and it remains visible until a day or two after full Moon, when the shadows begin to encroach across it. The exact days when you can see it depend very much on the *libration* of the Moon (see page 31).

Point Score

⬤⬤⬤⬤⬤⬤⬤⬤⬤⬤	Date seen	Points
Finding Mare Fecunditatis		1
Finding Messier and Messier A		2
Finding Langrenus		1
Finding Petavius and its trench		1 each
Score		

What you'll see

Despite its name, no life can be found on this lava plain. Like the rest of the Moon, it has been lifeless since its origin, and the name was given in the 17th century when the true nature of the Moon was not known. The surface is rippled by wrinkle ridges like those on **Mare Serenitatis**, best seen after full Moon. They look a bit like the skin of an old, shrunken apple. The weight of all the volcanic lava in the sea has squeezed its surface together.

Look out for one really curious feature – the twin craters Messier and Messier A.

▶ Part of Mare Fecunditatis with the twin craters Messier and Messier A and their strange rays.

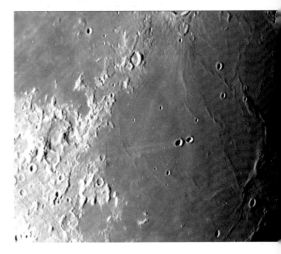

Unlike most craters of their size, about 10 km across, they are not completely circular in outline, but are quite elongated. Craters like this were formed billions of years ago when huge chunks of rock were hurtling around the Solar System in all directions, at speeds of many kilometers a second. Imagine a body maybe 500 m across hitting the Moon at this speed. In a fraction of a second it would penetrate some way into the surface before the force of the impact vaporized it.

The result – a huge explosion from below the surface that creates a circular crater no matter what direction the body came from. But in the case of the Messier pair, it looks as though the body struck a glancing blow, and even skipped like a flat stone on water. The debris from the impact formed two obvious bright rays to the west, which are very clear when the Sun is high over the sea, around full Moon.

On the western shore of Mare Fecunditatis is a giant crater, Langrenus. And to its south lies an even larger one, Petavius. This has mountain peaks near its center and a deep trench running from these peaks to the edge, making it look like a clock with one hand stuck at about 8 o'clock.

Fact File

Name	Diameter
Mare Fecunditatis	840 km
Messier	9 × 11 km
Langrenus	132 km
Petavius	177 km

◀ Petavius, photographed using a 70 mm refractor and a cheap camera.

4

Mare Tranquillitatis

This lunar sea is the site of the first manned Moon landing, back in 1969. So is it possible to spot the lunar module with your telescope?

Where to look

The "Sea of Tranquility" is the next dark marking in from **Mare Crisium**. Look for the little oval of Mare Crisium and the diamond shape of Palus Somni, then Mare Tranquillitatis is the dark patch adjacent to Palus Somni. It has a few pointed headlands jutting into the sea. Lunar seas are not true seas, but are regions where molten Moon rock solidified billions of years ago following giant impacts from space debris.

What you'll see

Mare Tranquillitatis has a mottled floor and is dented by a few craters. They may look small in your telescope or binoculars, but the largest, Plinius, on the northern side, is 43 km across. That is the size of a major city.

To the left-hand side of the sea as seen in binoculars are a pair of similar-sized craters, Sabine and Ritter, each about 30 km across. These two point the way to the site of Tranquility Base (American spelling with one "L" as it is an American

Point Score

●●●●●●●●●●●	Date seen	Points
Finding Mare Tranquillitatis		1
Finding Sabine and Ritter		1 each
Finding Cauchy		3
Finding Cauchy Rille and Scarp		4
Viewing Cauchy Rille and Scarp at both sunrise and sunset		5
Score		

◀ The famous Mare Tranquillitatis, site of the first lunar landing in 1969.

▼ Cauchy, with the Cauchy Rille above it and the Cauchy Scarp below it, photographed with a 200 mm telescope.

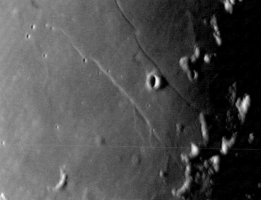

base), about 60 km away, There's nothing obvious to mark the spot as seen from Earth, however.

The problem is that the base of the Lunar Module, which is what remains on the Moon, is only 9 m across. If you compare that with Sabine or Ritter at 30 km across, you'll realize that it is far too tiny to be seen, and even the largest telescope on Earth will not show it.

Also on the Mare Tranquillitatis are two cracks in the Moon's surface called the Cauchy Rille and Scarp. Cauchy is a crater near Palus Somni, and on either side of it run streaks which can only be seen when the area is close to the terminator – the edge of the Moon's shadow line. This is either when the Moon is a crescent a few days after new Moon, or gibbous a few days after full Moon.

The northern streak is the Cauchy Rille. A rille is a type of valley where the land has dropped between two parallel faults. This one is over 200 km long. To the south of Cauchy itself is the

120 km scarp, which is a single fault resulting in a steep slope with the land higher on one side than the other. The scarp appears as a dark line after new Moon as it casts a shadow on the plain, after lunar sunrise, but as a bright line after full Moon as the scarp slope is catching the sunlight near sunset.

Our Moon map shows the landing sites of Apollo and other spacecraft that have visited the Moon.

Fact File

Name	Diameter	Name	Diameter
Mare Tranquillitatis	540 × 780 km	Maskelyne	24 km
Plinius	43 km	Sabine	30 km
Arago	26 km	Ritter	29 km
Ross	25 km	Cauchy	12 km

5

Mare Serenitatis

Looking at this lunar sea, you can understand why solid lava might be mistaken for water – it has distinct waves across it.

Where to look

Mare Serenitatis, the "Sea of Serenity," is one of the more-or-less oval dark patches on the Moon, and it's visible between just before first quarter until a few days after full Moon. One edge of it becomes visible six days after new Moon, then about a day before first quarter it is fully on show. This is the best time to see it, because then the features are seen under low illumination and they show up well. Its left-hand edge merges with the neighboring **Mare Imbrium**.

What you'll see

The sea itself is pretty plain, with just one significant crater, the 16 km Bessel. On the sea's eastern edge is a large and battered crater called Posidonius, around 100 km across. It lies on the edge of a sort of inlet from the main sea, called Lacus Somniorum, the "Lake of Dreams."

Not far from Posidonius, running across the sea itself, are some long, wriggly ridges that are easily seen just as the Sun has risen over them, but are less obvious when the Sun is higher. These are called *wrinkle ridges*, and they are found on many of the seas, but these are particularly prominent. At first glance you might think these are waves in a sea, but of course – unlike real waves – they never move.

Fact File

Name	Diameter	Name	Diameter
Mare Serenitatis	675 km	Aristoteles	87 km
Bessel	16 km	Hercules	69 km
Posidonius	100 km	Atlas	87 km
Eudoxus	67 km	Bürg	40 km

This is a good time to look for other craters in the area. To the north of Mare Serenitatis are two quite large ones, Eudoxus and Aristoteles, while farther east of these is another pair, Hercules and Atlas. Between these two pairs is a place where you wouldn't want to end up – Lacus Mortis, the "Lake of Death," with the crater Bürg close to its center.

In 1972, two lunar explorers landed near the southern edge of Mare Serenitatis on the Apollo 17 spacecraft. One of them, Harrison Schmitt, was a trained geologist, and he and Eugene Cernan spent three days driving around the immediate area aboard a Lunar Rover to study the lunar surface.

Point Score	Date seen	Points
Finding Mare Serenitatis		1
Finding Posidonius		1
Finding Bessel		1
Seeing the wrinkle ridges at lunar sunrise or sunset		3
Finding Eudoxus, Aristoteles, Atlas and Hercules		1 each
Score		

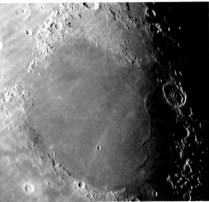

▲ Mare Serenitatis and its wrinkle ridges, with Posidonius on the right shore.

◄ On the surface of Mare Serenitatis. The crater at right, nicknamed Shorty, is 14 m deep and about 110 m across – too small to be seen from Earth. There are unusual patches of orange soil within it.

6

Theophilus, Cyrillus and Catharina

Three craters for the price of one object – that's the deal here! These adjoining craters make a distinctive sight.

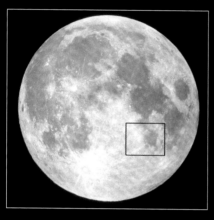

Where to look

When the Moon is a quite fat crescent, just a day or so before first quarter, you will see a small, roughly circular lunar sea to the west of **Mare Fecunditatis**, known as Mare Nectaris, the "Sea of Nectar." Just to the west of this is a trio of large craters, which are Theophilus, Cyrillus and Catharina. They remain visible for another two weeks, though they are much less easy to see when the Moon is around full, but they become easily visible again a few days before last quarter.

What you'll see

At the right time, when they are close to the Moon's terminator (the shadow line), the three craters stand out from the surroundings very clearly. They can even

make an obvious dent in the terminator, which keen-eyed people might spot without any optical aid at all, and are visible with binoculars. Theophilus is the clearest of the three, but together they make a nice group.

Also look for

Nearby, on the south shore of Mare Nectaris, is a giant ruined crater called

Point Score

●●●●●●●●●●●	Date seen	Points
Finding Theophilus		1
Finding Cyrillus		1
Finding Catharina		1
Finding Fracastorius		1
Finding Altai Scarp		2
Score		

Fracastorius. Billions of years ago this was an ordinary large crater, but then a great impact nearby, which created Mare Nectaris, melted the rocks and the lava flowed into it and solidified. Now we see it frozen in time, with the sea permanently flooding its interior.

There are other signs of the impact that created Mare Nectaris nearly 4 billion years ago. Look to the southwest of the sea and when the lighting is right you can see a long wriggly line, known as the Altai Scarp, shown as Rupes Altai on the map. Before first quarter it is bright, and after full Moon it is dark.

▼ Mare Nectaris, with Theophilus, Cyrillus and Catharina on its western shore. You can see the Altai Scarp wriggling to their lower left. This photo was taken before first quarter, so the scarp appears bright.

This is a curved fault line which runs some distance from the edge of Nectaris, and was caused by a cracking of the Moon's crust when the impact happened.

A scarp is a long inclined cliff, and you can tell which way this one faces by whether it is bright or dark. Its face points east, so it catches the sunlight before first quarter but is in shadow after full Moon. It's quite hard to spot when the sunlight is high over it, as it merges into the rest of the landscape.

Fact File

Name	Diameter	Length	Height
Theophilus	100 km	—	—
Cyrillus	98 km	—	—
Catharina	100 km	—	—
Fracastorius	124 km	—	—
Altai Scarp	—	425 km	Up to 1,000 m

7

Mare Imbrium

One of the most fascinating areas on the Moon, this lunar sea has flooded craters and drowned mountains.

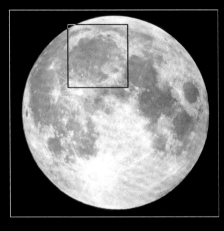

Where to look

Mare Imbrium covers much of the north-western half of the Moon, so it starts to become visible only after first quarter. It is slowly revealed as the Sun rises across it over a period of a few days, and can then be seen right up to last quarter when the Sun sets over its shores.

Point Score

	Date seen	Points
Finding Mare Imbrium		1
Finding Plato		1
Finding Alpine Valley		2
Finding Mare Frigoris		1
Finding Pico and Piton		2 each
Finding Aristillus, Autolycus, Archimedes, Timocharis and Lambert		1 each
Score		

See also: **Sinus Iridum.**

What you'll see

Its name means "Sea of Rains," but all that has rained on Mare Imbrium is space debris that has created additional craters on its surface. The sea itself is a sort of crater in its own right, as like all the other lunar seas it formed in an enormous collision between the Moon and a seriously large chunk of space rock billions of years ago. The force of the collision melted the rock and created a lake of molten lava, which flooded over 1,000 km of the lunar surface. Craters and mountains that were there previously were drowned in this catastrophic event, and you can still see their remains through a telescope.

Mare Imbrium was probably the last of the giant seas to be formed – but even so, it was created around 3.8 billion years ago. Astronomers can judge the age of a surface in the Solar System just by counting the number of craters on its

▶ A close-up of Plato, with the drowned mountains Pico and Piton, and the Alpine Valley. Part of Mare Frigoris is at the top.

surface, working on the basis that the older the surface, the more it will have been hit.

Begin your tour of Mare Imbrium at its northern edge. Just beyond the shore is another circular dark area, the crater Plato. This is one of the darkest features on the whole Moon. When the Sun is just rising over the crater, the jagged shadows of the west wall spreading across the crater floor are long, but even a couple of hours later you can see that they have shortened, and the next night they are shorter still. To the west of Plato is what looks like a gash through the mountains – called the Alpine Valley (Vallis Alpes), as these are the lunar Alps. This is where the land has sunk between two parallel rock faults.

North of Plato is another sea with shores that zigzag right across the northern edge of the Moon. This is Mare Frigoris, the "Sea of Cold" – though in fact, despite being in the far north, it receives as much sunlight as any other part of the Moon so the surface is searingly hot during the day.

Look on the sea floor near these features and you can see the peaks of ancient mountains sticking up through the solidified lava. One is called Pico and another is Piton. To their south are three large craters – Aristillus, Autolycus and Archimedes – with a range of mountains, the lunar Apennines, forming the southwest shore of the sea.

A few smaller craters, such as Timocharis and Lambert, dot the southern part of Mare Imbrium.

Fact File

Name	Diameter	Length	Height
Mare Imbrium	1,145 km	—	—
Plato	101 km	—	—
Alpine Valley	—	160 km	—
Mare Frigoris	—	1,440 km	—
Pico	—	—	2,400 m
Piton	—	—	2,300 m
Aristillus	55 km	—	—
Autolycus	39 km	—	—
Archimedes	83 km	—	—
Timocharis	34 km	—	—
Lambert	30 km	—	—

8

Sinus Iridum

Though not a major lunar feature, sunrise over Sinus Iridum is a dramatic sight for observers with telescopes.

Where to look

About three or four days after first quarter, this bay on the northwestern shore of **Mare Imbrium** starts to come into view. This is an event to catch when the Sun is low over the bay, but not every month will be just right for observing it, as one night the terminator might fall short of it and the next it could be beyond it. So you may have to wait quite a while before conditions are ideal. To catch sunset over Sinus Iridum, you would have to observe when the Moon is a crescent in the early hours of the morning. Its name means "Bay of Rainbows."

What you'll see

The bay is a semicircular bite out of the edge of Mare Imbrium. Binoculars will show it, though not very clearly, and a telescope with a moderate magnification – say 50 or 75 – is needed to get the most dramatic views.

Point Score

●●●●●●●●●	Date seen	Points
Finding Sinus Iridum		1
Sunrise over Sinus Iridum		5
Score		

The best experience comes when the Sun is starting to rise over the bay. The two extremities are marked by promontories, or headlands. The eastern one, Laplace, is the first to catch the sunlight, while the rest of the bay is still in shadow. Then the magic happens, as a few points of light appear beyond it in the darkness. Eventually the points join up, and the far wall of the bay starts to appear, even though the floor is still pitch black. Finally, the whole semicircle is visible, including the western promontory, Heraclides. The hills that form the semicircular edge to the bay are the Jura Mountains (Montes Jura).

The whole process takes many hours to unfold, so it is unlikely that you will see more than a part of it at any one time. You have to be lucky with your timing to catch the very first stage, when only a small part is visible. Then the next night, the shadow line has moved a lot farther across the Moon's surface and the bay takes on a fairly ordinary appearance. Each month, you are likely to see a different stage, weather permitting.

If you view Sinus Iridum in the early morning, as the Sun is setting over it, you will see the shadows of the Jura Mountains cast over the surface of the bay.

Brain Box

Notice that it's called Sinus Iridum – not Sinus Iridium! Even some people who should know better get this wrong.

Fact File

Name	Diameter	Height
Sinus Iridum	236 km	—
Jura Mountains	—	3,000–4,700 m

▼ One of the must-see sights of the Moon – sunrise over Sinus Iridum, photographed in this case using a 60 mm refractor.

9

Mare Nubium and the Straight Wall

This area of the Moon is home to one of its most famous features, the Straight Wall – but names can be misleading!

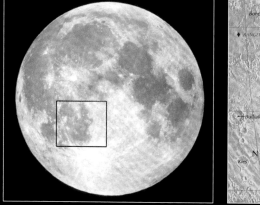

Where to look

This part of the Moon is visible from a day or so after first quarter until about last quarter, just slightly to the southwest of the center of the visible face. It's the most southerly part of the great gray plain that covers most of the western half of the Moon, and is to the southeast of **Oceanus Procellarum**. Mare Nubium means "Sea of Clouds" – a fanciful name, like all the other names of lunar "seas." The best time to look is when the terminator is quite close, so there are shadows nearby.

What you'll see

The most striking thing on the surface of Mare Nubium is the Straight Wall. (What! A wall on the Moon?) At first glance it really does look like a straight line on the surface of this sea, but when you look more closely it is not really straight. It is a scarp or escarpment, like the Altai Scarp (see page 39), and is a fault line where the land to the west is lower than that to the east. As it runs

Fact File

Name	Diameter	Length	Height
Mare Nubium	700 km	—	—
Straight Wall	—	110 km	Up to 400 m
Thebit	57 km	—	—
Birt	17 km	—	—
Arzachel	97 km	—	—
Alphonsus	108 km	—	—
Ptolemaeus	153 km	—	—

roughly north–south, the sunlight shines on the Wall after last quarter, so it appears bright, but before first quarter it is in shadow and appears as a black line.

What would you see if you could stand near the foot of it on the Moon? You wouldn't see any sort of a wall or cliff, but just a rather steep slope of about 20°, or about 40%, no worse than many hills on Earth, stretching as far as the eye could see in either direction.

Also look for

To the east of the Straight Wall is a rather neat crater named Thebit. This has a smaller crater exactly in the middle of one of its walls, and there's a smaller crater still on the edge of that. And to the west of the Wall is Birt, which also has a smaller crater breaking into its wall.

Just north of Thebit is a larger crater called Arzachel, with another even larger,

Point Score	Date seen	Points
Finding Mare Nubium		1
Finding Straight Wall after first quarter		2
Finding Straight Wall after full Moon		3
Finding Thebit		1
Finding Birt		1
Finding Arzachel, Alphonsus and Ptolemaeus		1 each
Score		

Alphonsus, north of that. Keep going and you come to a larger crater still, Ptolemaeus, which has a flat, plain floor. This chain of craters is easily recognizable and often helps you to get your bearings. When viewed with south at the top, as with most astronomical telescopes, they are sometimes referred to as the "Lunar Snowman," being a bit like a snowman made from three different-sized balls of snow.

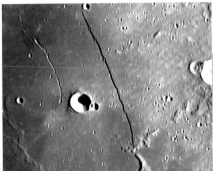

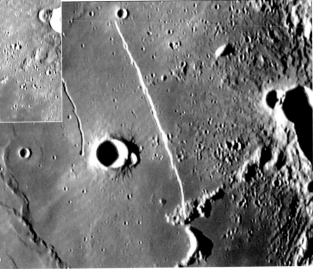

▲▶ The Straight Wall appears as a dark line at first quarter (above) and as a bright line at last quarter (right). It is shown on the map as Rupes Recta. The crater to its left is Birt, with Thebit at the right edge of the picture.

45

These two nearby prominent craters have the ability to change their appearance completely, depending on when you view them.

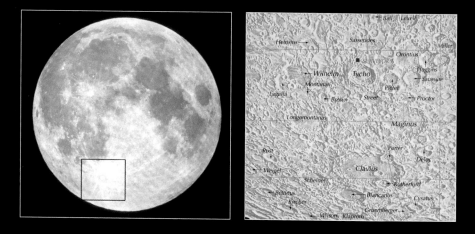

Where to look

Tycho and Clavius are both toward the Moon's south pole, and are in its western hemisphere. They become visible a couple of days after first quarter, and disappear into shadow again after last quarter.

But while most craters, including Clavius, merge into the pattern of light and dark around full Moon, Tycho is at its most obvious then, becoming very bright and at the center of a huge pattern of rays or spokes. These look quite like the grid lines on a map of the polar regions, all spreading out from one spot, though Tycho is not at the south pole itself.

What you'll see

Tycho is not a particularly large crater, but it stands out from all the others in this area because of its sharpness. Of all the large lunar craters, Tycho is one of

the youngest, dating from only about 180 million years ago, which is about the same as **Aristarchus**, though these dates are difficult to work out and may be wrong by hundreds of millions of years. All the nearby craters have rounded surfaces and have been battered by billions of years of minor impacts. And the closer you get to full Moon, the more obvious it becomes, as it is very bright.

The rays that radiate from Tycho extend for as much as 1,500 km across the lunar surface, and are made of material that was

Point Score ⚫⚫⚫⚫⚫⚫⚫⚫⚫	Date seen	Points
Finding Tycho		1
Seeing Tycho's rays		1
Finding Clavius		1
Score		

thrown out of the site by the giant impact that caused the crater. Look for the darker zone around it, which was too close for this material to land.

Clavius is a much older and larger crater – in fact, it's one of the largest on the Moon. It is so big that if you were standing in the center, the curvature of the Moon would mean that you would not see the crater walls, over 100 km away. Clavius has a very distinctive arc of smaller craters on its floor, together with the larger crater Rutherfurd on its rim. They are so regularly placed, and in decreasing order of size, that you might imagine that they were all created together, but their positioning and sizes are probably just a matter of luck.

Fact File

Name	Diameter
Tycho	85 km
Clavius	225 km
Rutherfurd	48 km
Clavius D	28 km
Clavius C	21 km
Clavius N	13 km
Clavius J	12 km

Brain Box

In times gone by, when there were no streetlights, meetings and other events would be organized around full Moon so that people could see their way home afterward.

▶ Tycho is at the top left of this photo, with Clavius and its row of baby craters just below the center. After Rutherfurd, the craters are lettered, in order of decreasing size: D, C, N and J.

II

Oceanus Procellarum

The Moon's largest sea is studded with fascinating craters that are some of the most dramatic sights in a telescope.

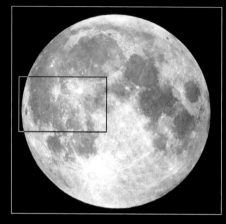

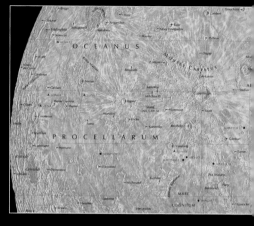

Where to look

A couple of days after first quarter Oceanus Procellarum starts to come into view, and it's visible in part until the Moon becomes a thin crescent again. It occupies virtually all of the northwestern part of the Moon – an enormous gray plain that is anything but featureless. Because it is so large, you need to observe it over a period of several days so that you view its features under varying illuminations. You can see the area easily with the naked eye, but binoculars will show some of the features and a small telescope will show them very well. Its name means "Ocean of Storms" and it is the Moon's only Ocean – everything else is a Sea or a Bay or something else.

What you'll see

The rather flat gray plain itself is not particularly exciting, but the craters that cover its surface more than make up for it. Unlike what are called the highland areas in the southern part of the Moon, which are lighter in color and are just a mass of old, rounded craters that have been battered to bits by incoming space debris over billions of years, the craters in Oceanus Procellarum were formed rather later in the Moon's

Point Score		
⬤⬤⬤⬤⬤⬤⬤⬤⬤⬤	Date seen	Points
Finding Oceanus Procellarum		1
Finding Copernicus		1
Finding Kepler		1
Score		

◄ Copernicus is a magnificent crater, and you can spend a long time studying its ramparts when they are side-lit like this. The hills to its north are called the Carpathian Mountains.

Fact File

Name	Diameter
Oceanus Procellarum	2,500 km
Copernicus	93 km
Eratosthenes	58 km
Stadius	69 km
Kepler	32 km

history. Some of the lava flows on the surface are as young as a billion years or so – which on the Moon is quite recent! So the craters must be younger than that.

The most obvious crater in the whole area is Copernicus. The impact that created it probably took place about 800 million years ago, at a time when the only creatures on Earth were things like algae and tiny protozoa in the oceans – no dinosaurs or people around then to see it happen, though what a sight it would have been! Yet it looks quite sharp and fresh, and there are plenty of signs of the impact all round it which you can see with a telescope – lighter rays spreading out from it, and chains of smaller craters best seen just after the Sun has risen over them.

Nearby to the east is a smaller crater, Eratosthenes, which is on the shores of an area called Sinus Aestuum, the "Bay of Billows." In places there are older craters that have been drowned by lava flows. Stadius, between Copernicus and Eratosthenes, is an example.

The crater Kepler, to the west of Copernicus, and visible a couple of days later, is smaller and lacks a central peak. Kepler and Copernicus are easily visible when the Sun is high over Oceanus Procellarum because of the bright rays around them. In fact, if you have sharp sight you can try looking for these with the naked eye.

Also look for

Also within Oceanus Procellarum is a very interesting crater, **Aristarchus**, visible as a very bright spot with binoculars. Right on the western edge of the Moon, beyond Oceanus Procellarum, is an oval dark feature called Grimaldi, 430 km across. This, like **Mare Crisium** on the opposite edge of the Moon, shows the effects of *libration* very clearly, sometimes being close to the limb and sometimes farther in.

The Moon's brightest feature is in an interesting area of the Moon that is fun to watch under different illuminations.

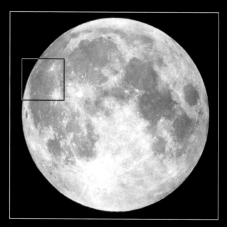

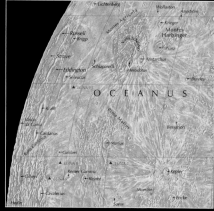

Where to look

Aristarchus is quite a small crater in **Oceanus Procellarum**, but it is usually very easy to find as long as it is in sunlight because it is so bright. It is in the northwestern part of Oceanus Procellarum, and is visible from about three or four days before full Moon until the Moon is a crescent in the early morning sky.

What you'll see

Aristarchus is quite a young crater, at only around 170 million years. By that time, Earth was teeming with life, with dinosaurs and the first mammals roaming the forests, but of course no people were around to witness the impact on the Moon that caused it. Most of the other features on the Moon are billions of years old. Lunar rock that has been churned up becomes darkened by what's called space weathering – bombardment by particles from the Sun and by tiny meteorites – so we know that Aristarchus is comparatively young.

Aristarchus can be seen with binoculars, and it's so bright that you can sometimes even spot it when that part of the Moon is faintly visible beside a thin evening crescent Moon (see the item on Earthshine on page 29). But for good views, you need a telescope with a magnification of maybe 100 times.

Point Score	Date seen	Points
Finding Aristarchus		1
Finding Herodotus		1
Finding Schroeter Valley		2
Score		

Look with a telescope to the west of Aristarchus and you will see what at first glance looks like a tadpole, with a large head and a wriggly tail. The head is the crater Herodotus, but the tail is not actually connected to it. It is known as the Schroeter Valley, which starts with a crater and then wriggles away for 160 km across the lunar surface. It is often called the Cobra Head. This is an example of a sinuous rille, or lava channel, with the lava emerging from the crater at the head of the valley.

To the northeast of Aristarchus is a range of hills known as the Harbinger Mountains, so-called because they appear before Aristarchus as the sunlight progresses across the Moon, so they are harbingers of dawn on Aristarchus itself.

South of Aristarchus is another range of hills, the Marius Hills. These are volcanic domes, where lava has forced the surface upward.

▼ Aristarchus is the brightest feature on the Moon, but if you look at it under high magnification you can see darker rays within it.

Fact File

Name	Diameter
Aristarchus	40 km
Herodotus	35 km

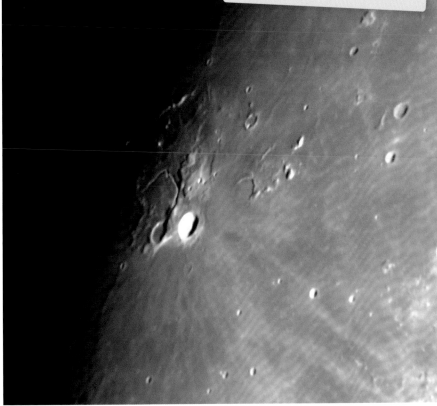

You have to be quick to catch elusive Mercury – no sooner has it appeared in the sky than it's gone again.

▶ Mercury usually appears like a bright star in the twilight. But if you could see its orbit as well, this is how it would look. At the time, Mercury was 5° above the horizon and 15° from the Sun.

Where to look

You'd think Mercury should be easy to observe. It appears at sociable hours in the evening sky, so you don't even need to lose any sleep in observing it. And it's the closest planet to the Sun, so it's up there with the brightest planets. But this is also the reason why it's so hard to find – it only ever appears in a bright sky. There's just a short period between the sky being too bright after sunset or before sunrise and the planet being too low in the sky for it to be observed.

Point Score

●●●●●●●●●●	Date seen	Points
Seeing Mercury		3
Seeing the half phase		5
Score		

As Mercury swings around the Sun, about every two months it moves far enough out that it peeps above the horizon in either the evening or the morning twilight. But there is a catch – even when it is at its greatest distance from the Sun (known as its *greatest elongation*), Mercury can often be low in the sky because the line of its orbit in the sky is close to the horizon. As a result, if you want to see Mercury in the evening sky, it's best to look when it appears in the spring, or in the morning sky in the fall. So the result is that there are normally only one or two elongations each year, as shown in the list on the facing page, that are any good.

Viewing dates

The best dates to see
Mercury in the evening
sky from the northern
hemisphere. It is highest in
the years when the elongation
occurs in March or April.

2015	January 15
2015	May 7
2015	December 29
2016	April 19
2017	April 1
2018	March 16
2019	February 27
2019	June 24
2020	February 11
2020	June 5
2021	January 24
2021	May 17
2022	January 7
2022	April 29
2022	December 22
2023	April 11
2024	March 4
2025	March 8

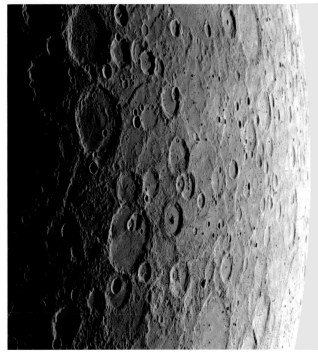

▲ Close up, Mercury looks very much like the Moon, with masses of craters, but with few seas or lava flows. This shot was taken by the NASA Messenger spacecraft in 2012.

You can start to see Mercury from about half an hour after sunset for a week or so on either side of these dates, for up to 45 minutes before it gets too low. Look in the western twilight, just to the left of where the Sun has set. If there is a crescent Moon or another planet visible, that will help to show the line that the Moon and planets take through the sky.

What you'll see

Because Mercury is not much bigger than the Moon, but is about 340 times farther away on average when it is at greatest elongation, it is only very small even when seen through a telescope. At that time we see it with the Sun coming from the side, so it looks like a tiny half Moon, though you would need a magnification of about 250 to see it in a telescope looking the same size as the Moon does with the naked eye.

Mercury has no atmosphere, and spacecraft photos show that it has a cratered surface, like a larger version of our own Moon.

Fact File

Diameter	4,878 km
Typical distance from Sun	57.9 million km
Mass compared with Earth	1/18th
Typical surface temperature	167°C
Moons	0

The brightest planet of all, and our nearest planetary neighbor, beautiful Venus can still present a challenge.

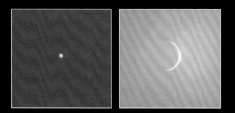

▲ Two views of Venus through an 80 mm refractor. On the left is the distant globe when it's on the far side of its orbit from Earth, in this case 233 million km away. On the right is the way it appears when it's on the near side of its orbit, as a thin crescent, just 42 million km away.

▲ A few Soviet spacecraft landed on Venus in the last century and sent back views of the surface for a short while, until they failed due to the intense heat and surface pressure. This view from Venera 13, taken in 1982, shows flat rocks and part of the spacecraft.

Where to look

When Venus is around, everyone knows about it, as it outshines everything in the night sky apart from the Moon. Like Mercury, it appears only in the evening or morning sky, either following the Sun down in the west after sunset or rising before it in the east before sunrise. But it can get much farther from the Sun than Mercury, as it is in a larger orbit around the Sun. So it remains in the sky for months at a time rather than just days, and can appear quite high in the night sky rather than hiding in the twilight.

But for the same reason that Mercury is often hard to see even when at its greatest elongation from the Sun, some years are better for spotting Venus than others. Years when the evening elongation occurs in the spring, such as 2020, provide the best viewing opportunities, with the planet really high in the sky. In some other years, such as 2018, however, Venus will be hard to see from northern states and Canada.

What you'll see

Because Venus is closer to the Sun than Earth is, we can see it going through a range of phases like those of the Moon. When it is on the far side of its orbit it is illuminated by the Sun almost full on, but when its orbit brings it closer to Earth it appears as a very thin crescent. So during

Point Score	Date seen	Points
Seeing Venus		1
Seeing the half phase		2
Seeing the crescent phase		1
Score		

54

an evening viewing season of the planet, it first appears close to the horizon as a tiny circular dot, then it grows larger and gets higher in the sky until at greatest elongation it is at half phase. Then it continues to grow and becomes a thin crescent, though by now it is getting low in the sky again. You can see the phase at this time very clearly even with binoculars. The morning sky appearances are the reverse of the evening ones. The planet starts off as a large, thin crescent and ends up months later as a tiny dot.

But seeing anything other than the phase of Venus is another matter. The planet is completely cloud-covered, so all it shows are a few elusive dusky markings from time to time. Beneath the white clouds – which are sulfuric acid rather than water – lies a very hot planet with no trace of water and a barren volcanic landscape. Conditions on the surface are very hostile to human life, and it will probably be a long time before anyone goes there.

Fact File

Diameter	12,104 km
Typical distance from Sun	108.2 million km
Mass compared with Earth	4/5th
Typical surface temperature	464°C
Year length	225 Earth days
Day length	243 days
Moons	0

Observing tips

To get the best views, try to observe the planet while it is still in the twilight sky, as soon as it becomes visible. It can be hard to spot, and even Go To telescope owners have trouble as there are usually no stars visible in the sky to help in setting up the telescope. So the easiest way to find the planet is to catch it on those occasions when the Moon is close to it. The Moon is easy to spot in a clear sky, and you should then be able to see Venus nearby. Computer sky programs or mobile phone apps will show you the best times each month.

Viewing dates

The best years to see Venus in the evening sky from the northern hemisphere and the date of greatest elongation.

2015	June 7
2017	January 13
2020	March 25
2023	June 4
2025	January 10

▶ The clouds of Venus are so dense that the only way to map the surface is by radar. An orbiting American spacecraft called Magellan did this in the 1990s. This perspective view shows lava flows and volcanoes.

Everyone has heard of the Red Planet, but it isn't always easy to see, despite being one of the closest planets.

▶ Through an 80 mm refractor you can catch a glimpse of a polar cap, dark markings and some atmospheric clouds when the planet is at opposition.

Where to look

Mars is in the next orbit out from Earth, and it moves more slowly in its path around the Sun. This means that we can catch it up in its orbit, then after just a few months we have gone past and it's another two years before we catch it up again.

The times when we see it best are called *oppositions*, a term which applies to all the planets farther out from the Sun than Earth. So when you hear about the opposition of Mars or Jupiter, it doesn't mean that the planet is being awkward or argumentative, just that it is opposite the Sun in the sky.

Oppositions of Mars happen at intervals of just over two years, but for long periods of time the planet is very distant and difficult to observe. And because the orbit of Mars around the Sun is not very circular, its distance at each opposition varies so there are close and distant oppositions. Close oppositions occur in 2018 and 2020. In 2018 the planet is very low in the sky as seen from North America, so the 2020 one will be easier to observe.

The dates of the oppositions of Mars are shown in the list here, which also tells you the constellation where you can find it on the date of opposition. But because Mars moves quickly, it will be in that constellation only around that date. Earlier in the year it will be a constellation east, and later in the year it will be a constellation west.

Fact File

Diameter	6,792 km
Typical distance from Sun	227.9 million km
Mass compared with Earth	1/9th
Typical surface temperature	−55°C
Year length	687 Earth days
Day length	24h 37m
Moons	2

What you'll see

Although it's called the Red Planet, don't take this too literally and expect to see it looking like a traffic light. Without a telescope it looks more pink than red, though when it is really close and bright the color is more obvious. Sometimes, when it's really close, it is the brightest planet after Venus, but for most of the time it's not particularly dazzling.

Mars is quite a small planet – only about twice the diameter of the Moon. Even when it's at its closest it appears about 75 times smaller than the Moon, and usually it's much smaller than that. You need a telescope to see anything on it at all, and for most of the time all you can make out is a tiny orangey disk. If you are lucky you might see one of its white polar caps and a dark marking or two. These are not as prominent as photos of the planet make out, though, so if you don't see anything straight away, keep at it.

Point Score

⦿⦿⦿⦿⦿⦿⦿⦿⦿⦿	Date seen	Points
Seeing Mars		2
Seeing a polar cap or markings on Mars		5
Score		

A small target like Mars is affected by what astronomers call *seeing* – that is, the turbulence of our atmosphere. Sometimes, when the seeing is bad, the disk is all over the place and never stays still. When the seeing steadies down, you can get moments where everything is sharp, but even then the markings can be elusive. The seeing gets better the higher the planet is in the sky, and you can often

▼ The Martian surface is bitterly cold, bleak and rocky, but the planet does have a thin atmosphere which makes it look the most Earthlike of all the other planets. This picture was taken by the Curiosity rover in 2012.

◀ A photograph of Mars taken through a large amateur telescope shows a lot of detail. South is at the top. The dark marking in the center of the disk is called Syrtis Major. The south polar cap has almost melted, but some blue clouds are hanging over the morning terminator and over the north polar regions.

Oppositions of Mars	
Date	Constellation
May 2016	Scorpius
July 2018	Capricornus
October 2020	Pisces
December 2022	Taurus
January 2025	Gemini

see more when the planet is high up but not as close, rather than when it is close but low down.

Observing tips

Use the highest magnification that still gives a fairly sharp image, and look carefully rather than trying to overmagnify what you see. A magnification of about 75 should be enough to show some details when the planet is close.

One side of Mars does not have strong markings anyway, so you might just be unlucky. Because Mars rotates on its axis in almost the same time as the Earth, if you observe at the same time the next night you will get pretty much the same view. So you either have to wait for several hours, or give it a week or so for another face of Mars to rotate into view.

Opposition and apparitions

When a planet is at opposition, we see it opposite the Sun in the sky, so it is best visible at midnight. This is usually within a few days of its closest distance to Earth, so it's the time when we get the best view of it. All planets farther out from the Sun reach opposition every so often, most doing so every year but Mars taking more than two years.

The period over which a planet is visible in our sky, and is not hidden by the Sun, is called an apparition. Each apparition begins with the planet emerging from behind the Sun and being visible in the early morning sky just before sunrise. At this time it is on the far side of its orbit and is very hard to observe. Slowly it rises earlier and earlier, getting farther from the Sun all the time, and eventually it starts to rise in the late evening.

At opposition it is due south at midnight, but you have to be up late to see it at its best. Then it becomes more easily visible in the evening sky, though getting more distant again, until eventually it starts to set not long after the Sun and soon becomes lost in the evening twilight.

This giant planet is the easiest of all the planets to observe, and comes round every year. Even with a small telescope there's something to see.

◄ Through a medium-sized telescope you can see the two major belts of Jupiter plus finer details which change from week to week.

Where to look

Jupiter is a very predictable planet. Not only is it really bright, so that it's obvious when it's around, but it repeats its performance year after year. Just check the dates of opposition, and it will be easy to spot in those months. Then the next year it has moved just one constellation to the east, so it is at opposition a month or so later. This is why there is no opposition in 2025.

What you'll see

Jupiter appears in the sky as a bright and creamy white object. You can't mistake it for a star because it doesn't twinkle. Only stars twinkle because they are points of light and their beam is easily disturbed by our atmosphere.

Even through binoculars you can see that there is something different about Jupiter. Though typical small binoculars don't show its disk, you can usually see what look like up to four tiny stars on either side of the brilliant dot that is the planet itself. These are its brightest moons, and they were spotted over 400 years ago by the famous Italian astronomer Galileo, who was one of the first people to point a telescope at the night sky. There's more about them on page 61.

You don't need much magnification on a telescope to show the disk of Jupiter. About 20 times will do, though 40 or 50 is better. To begin with you will just see the disk. Notice that it isn't a circular disk like the full Moon, but looks slightly fatter. This is because Jupiter spins very quickly, in under 10 hours. And because Jupiter is made of gas, not rock, the planet actually bulges as it spins.

Point Score	Date seen	Points
●●●●●●●●●●		
Seeing Jupiter		1
Seeing the flattened disk		2
Seeing the dark belts		2
Score		

◄ NASA's Cassini spacecraft flew past Jupiter in 2000, on its way to Saturn. This picture from it shows the amazing swirls of clouds in the planet's outer layers, and the huge Great Red Spot, big enough to swallow up three Earth-sized planets. The black dot is the shadow of one of its moons.

Oppositions of Jupiter	
Date	Constellation
February 2015	Cancer
March 2016	Leo
April 2017	Virgo
May 2018	Libra
June 2019	Ophiuchus
July 2020	Sagittarius
August 2021	Capricornus
September 2022	Pisces
November 2023	Aries
December 2024	Taurus
January 2026	Gemini

Look more carefully and you should spot darker bands crossing the fattest part of the planet. These aren't as dark as most photos show, but keep looking and you should see more and more fine detail. There are spots, such as the Great Red Spot, which at one time really was quite noticeable and red but now seems to be shrinking and losing its color.

Jupiter's features are not solid but are really swirls in the upper layers of clouds of gas. There is no solid surface that you could land on, and if you were to descend into the planet the gas would just get denser and hotter and at greater pressure until your spacecraft was crushed and melted.

Despite being made of gas, Jupiter is a huge planet and its gravitational pull affects the other planets and objects in the Solar System. Sometimes a comet or asteroid hits the planet, creating an impact scar that may last for weeks.

Fact File

Diameter (equatorial)	142,984 km
Typical distance from Sun	778.4 million km
Mass compared with Earth	318 times
Typical cloud-top temperature	−110°C
Year length	11.86 Earth years
Day length	9h 51m
Moons	At least 67

The four big moons of Jupiter are easy to see, even with binoculars. You can follow their movements night by night.

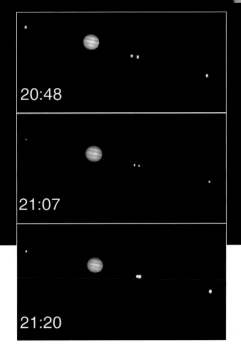

20:48

21:07

21:20

▶ Sometimes you can detect the movements of Jupiter's moons over a matter of half an hour or so. These are easiest to see when the fastest-moving moon, Io, is near another one so the separation between them changes quickly.

Where to look

Once you've found **Jupiter**, its four major moons are always there. Just look at the planet through any optical aid, such as binoculars, and you will see them, strung out on either side like beads on a necklace.

What you'll see

The four largest moons (or satellites) of Jupiter are quite bright, and if it were not for the glare of Jupiter itself you would be able to spot them even without binoculars on a good night. But although all but one of them are larger than our own Moon, as they are at the distance of Jupiter you need

a very large telescope and really steady conditions to see anything other than a dot of light. So we have to be content to watch them orbiting the planet night after night.

In order outward from Jupiter they are Io, Europa, Ganymede and Callisto. You can remember the order by thinking of the phrase "I Eat Green Cheese" (or Gorgonzola if you prefer). The closer the moon is to Jupiter, the faster it orbits the planet – which is like the Solar System in miniature, as the closest planet to the Sun orbits the fastest. Io takes less than two days for one orbit, so the chances are that every night you look it will be on the other side from the previous night. Callisto takes over two weeks for one orbit, and moves a long way out from the planet. However,

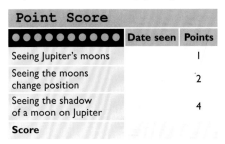

Point Score

●●●●●●●●●●	Date seen	Points
Seeing Jupiter's moons		1
Seeing the moons change position		2
Seeing the shadow of a moon on Jupiter		4
Score		

Fact File

Name	Diameter	Period of orbit around Jupiter
Io	3,630 km	1d 18h
Europa	3,138 km	3d 13h
Ganymede	5,262 km	7d 4h
Callisto	4,800 km	16d 17h

you can't assume that the most distant is always Callisto – it could just happen that it is almost in line with Jupiter so it appears closer than Io.

The satellites can move so quickly that you can almost see them move through the telescope. When one is close to another moon, you can see changes by looking every 10 minutes or so, if the two are moving in different directions. The same happens when one is moving on to or away from the limb of Jupiter. It looks as if it has some sort of growth! And a moon can suddenly appear or disappear, if it goes into Jupiter's shadow.

Sometimes you will see only three, two or even one moon. The others will be either behind or in front of the planet, in which case they merge in with the detail and are hard to spot. But quite often you can see the shadow of the moon on the planet, even with a small telescope. Only very occasionally are all four moons invisible. The next time this takes place and is visible from North America is in 2033! You can find out which moon is which online from www.calsky.com (go to Planets, Jupiter, Apparent View/Data).

All the other moons of Jupiter (more than 60 of them!) are too faint to be seen without a large telescope.

▼ An identity parade of Jupiter's Galilean satellites. Far left is Io, a mean-looking creature, with spots all over it caused by eruptions. Being so close to Jupiter, its innards get pulled around by Jupiter's gravity, causing it to heat up. Next up is Europa, an ice world whose surface appears to be a frozen ocean. But below its icy exterior could be a warmer ocean – and who knows what secrets it's hiding, maybe even primitive life! Then comes Ganymede, the Boss, because it's bigger than any other moon in the Solar System, even bigger than Mercury. Finally, at far right is Callisto, whose icy surface has been battered over billions of years by impacts and is more cratered than any other body in the Solar System.

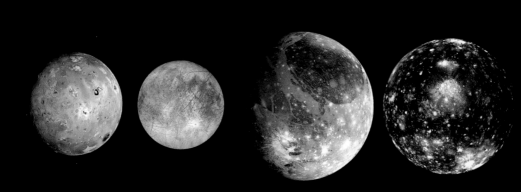

If there were a prize for the "Most Popular Planet," Saturn would win it every time. And you don't need a giant telescope to see it.

◄ Saturn appears as a bright star, though not twinkling. It's just above center in this photo, taken when it was in Cetus. It is always fainter than Jupiter and Venus, but not reddish like Mars.

Where to look

Saturn is the slowest mover of all the bright planets, so it's visible in more or less the same part of the sky from year to year. It will be in the summer skies for many years to come. It takes nearly 30 years to go round the Sun once, which means that it remains in the same constellation for two or three years at a time. The list of dates of opposition on page 65 tells you what time of year to look and in which constellation, but it'll be around in the evening sky for several months after those dates.

What you'll see

To the naked eye, Saturn is a bright star, but like the other planets this one doesn't twinkle. When it's at its closest to Earth – around opposition – Saturn can be brighter than most other stars visible from North America. But when it's more distant, it can be about half as bright, so then you might mistake it for a star. It looks distinctly yellowish when compared with most stars, which is often enough to pick it out as something special.

Viewing Saturn

As a planet, and ignoring the famous rings for the time being, Saturn has a lot less going for it than either Jupiter or Mars. But one thing is different – its globe is even more flattened than Jupiter's. This is because the planet is not solid but gaseous, and in fact its density is even lower than that of Jupiter. Overall, the planet is even less dense than water! Result – as it spins on its axis, the outer layers bulge noticeably.

Point Score ●●●●●●●●●●	Date seen	Points
Seeing Saturn		1
Seeing the flattened globe		2
Seeing belts		5
Score		

See also: **Saturn's rings**; **Titan.**

You can see this in small telescopes with a magnification of around 30 or so, though binoculars are not powerful enough to show the disk properly.

Like Jupiter, Saturn has dark belts and light zones, and you can see these with a telescope as small as 75 mm if you're lucky. But if you find it tricky to see detail on Jupiter, you'll really struggle with Saturn because the belts are rather less obvious. It depends on how good a view you get and your experience. With a small telescope on a poor night, when the planet is indistinct, you'll be lucky to see anything other than a pale yellow disk and blurry rings, particularly if you're new to the game. But give it a chance, and when conditions improve you should see detail even with a small telescope.

These days, amateur astronomers can observe more detail than ever before using CCD cameras, particularly using large telescopes. In particular, there are color changes that were not really noticeable in years gone by. Saturn's axis is tilted at an angle to the ecliptic, which means that it has seasons, just as Earth does. But there are no leaves to change color and fall in autumn on Saturn. Instead, the whole hemisphere changes color. Instead of being yellowish, the hemisphere that is tilted away from the Sun turns blue. You would turn blue if you were on Saturn, with temperatures around −140°C.

Amateur imagers have also discovered more storms in Saturn's clouds than were ever seen previously. Whether this is because Saturn is really becoming more lively or because we are just seeing what's there better now, it's too soon to tell.

In 2010, a major storm outbreak was seen on Saturn by amateur astronomers

▼ A Hubble Space Telescope view of Saturn as it appeared in 2004.

Oppositions of Saturn	
Date	Constellation
May 2015	Libra
June 2016	Ophiuchus
June 2017	Ophiuchus
June 2018	Sagittarius
July 2019	Sagittarius
July 2020	Sagittarius
August 2021	Capricornus
August 2022	Capricornus
August 2023	Aquarius

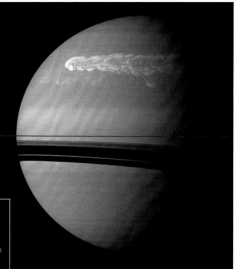

▲ A giant storm on Saturn, seen by the orbiting Cassini spacecraft in 2011.

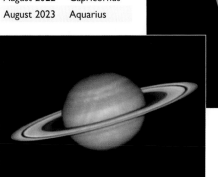

◀ Amateur observers were able to follow the same storm from Earth. This view was taken from Selsey, UK, using a 355 mm telescope.

even before it was picked up by the Cassini spacecraft in orbit around the planet. But Cassini was able to get some great pictures of the storm, which was larger than any on the planet since 1933 and lasted for months. At its height, the storm, which appeared in telescopes as a white spot, was visible even with quite small instruments.

Observing tips

Although Saturn's large and bright enough to be visible whenever it's in the sky, for the best views you really need to observe when it's at its highest in the sky, which means around opposition. But because Saturn takes so long to orbit the Sun, if you are unlucky you could have a very long wait for it to get really high in the sky. It spends half

its time below the celestial equator, which means that from the northern hemisphere it never gets high in the sky. This applies until 2026. So choose the time around opposition, when Saturn is due south at midnight, and just make sure you don't have to be up early the next morning! Bear in mind that the oppositions of Saturn all occur in summer, so astronomical midnight actually occurs around 1 am.

Fact File

Diameter (equatorial)	120,536 km
Typical distance from Sun	1,427 million km
Mass compared with Earth	95.2 times
Typical surface temperature	−140°C
Year length	29.46 Earth years
Day length	10h 13m 59s
Surface gravity	1.16 that of Earth
Moons	62

Everyone wants to see the rings of Saturn. And the good news is that they are actually quite easy to view, even with small telescopes.

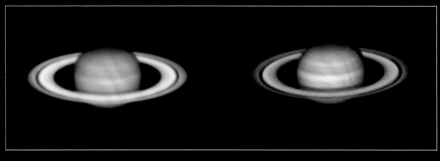

▲ When Saturn is at opposition, at left, the rings are much brighter than they appear two weeks later. These photos were taken through a 280 mm amateur telescope.

Where to look

Look at the previous page to find out where **Saturn** is. With the naked eye it looks the same as any other planet (see the article on **Jupiter** on page 59 to find out why it doesn't twinkle), but with any kind of optical aid you should start to see the rings.

What you'll see

Even binoculars with a magnification of 10 or more will show you that Saturn is not just a point of light but is slightly elongated, and once you get to a magnification of 20, with a small telescope, you can start to see the rings as well as the globe of Saturn. But if you didn't know what you were looking at, you might be puzzled as to why this planet appears different to Jupiter.

With a magnification of 30 or more you can really make out that the rings are separate from the globe – and it's a sight that even crusty old experienced astronomers never get tired of.

We say "rings" rather than "ring" because there really are several, but you need a medium-sized telescope – say larger than 75 mm aperture – to see more than one. The main rings are called A and B, with B being the brighter and wider, and the closer to the planet. A black space called the Cassini Division (after the astronomer who first spotted it) separates the two. Look carefully for the black space between the two – you'll need a telescope of 75 mm or larger to see it.

Point Score

⬤⬤⬤⬤⬤⬤⬤⬤⬤⬤	Date seen	Points
Seeing the rings		2
Seeing Ring A and Ring B separately		3
Seeing the Cassini Division		3
Score		

There is also a C ring, closer to the planet, sometimes called the Crepe Ring. You need a larger telescope to see this well, plus a very clear night and steady conditions, because it is rather fainter than the other two rings. These days, when we hear the word "crepe" we think of brightly colored crepe paper, or pancakes, but in Victorian times it meant a kind of gauzy silk, often black and worn by widows (also spelled "crape").

Each year, Saturn reaches opposition (see box on page 58), and something interesting happens – the rings become brighter. The rings are not solid but are made from tiny ice particles, and at opposition, when the Sun is directly behind us as we look at the planet, the particles behave a bit like those reflective road signs and shine more light directly back to us. This is the best time to view the rings, but it only lasts a week or so on either side of opposition.

The angle of tilt of the rings changes over Saturn's 29-year orbit of the Sun. At times it is wide open, and at others

▲ How the tilt of Saturn's rings changed between 2002 (top) and 2008, as seen from Earth.

it is almost in line with Earth. Around those times, the rings almost disappear and it's hard to see any detail at all in the rings, because they are so thin. This next happens around 2025.

Fact File

Diameter of rings	274,000 km
Thickness of rings	1 km
Composition	Water ice

▼ Seen backlit by the Sun from the Cassini orbiter, Saturn's rings are a spectacular sight.

Titan is one of the largest moons in the Solar System, and even though it is at the distance of **Saturn** it is easily visible with a small telescope.

◀ A large amateur telescope shows Saturn with several of its satellites. Titan is at bottom left, with Mimas, Tethys, Rhea, Enceladus and Dione from left to right closer to the planet.

Where to look

After you've found Saturn in a telescope, look near it for a starlike object. This could be a few planet diameters from Saturn itself, on one side or the other, or it could be nearer and above or below. The chances are that this is Titan, its largest moon, and the second largest moon in the whole Solar System after Jupiter's Ganymede.

It may be that you have seen an actual star, but check an hour or two later and a star will have moved noticeably (because Saturn moves in its orbit around the Sun) but Titan will be in almost the same spot.

Point Score

●●●●●●●●●●	Date seen	Points
Seeing Titan		2
Seeing Rhea, Tethys, Dione and Iapetus		4 each
Score		

What you'll see

Because Titan is quite distant you won't see its disk unless you have a very large telescope, and it just appears as a speck of light. It takes 16 days to orbit Saturn so if you look again the next night it will have moved only a short distance. But although it doesn't look very exciting through the telescope, Titan is a really interesting moon. Close-up pictures taken with the Cassini spacecraft and the Huygens lander, which it released in 2004, show an amazing surface. Titan has rivers and lakes, and an atmosphere of nitrogen.

On the face of it, this sounds like Earth – but there's one big difference. Titan's surface temperature is −179°C. So the lakes and rivers are not made of water, which is frozen, but methane. On Earth, methane is natural gas, which we cook with and run our gas central heating with,

◄ From Cassini, Titan is just an orange globe with its surface hidden within the smoggy nitrogen atmosphere.

but on Titan it is liquid. There may even be some sort of primitive life living in the lakes, but it will be decades before we find out for sure.

Fact File

Diameter	5,152 km
Period of orbit around Jupiter	15d 23h

Also look for

Unlike Jupiter's lesser moons, which are all quite faint, Saturn has some other moons which are just a bit fainter than Titan and which you might see through a telescope. The brightest is Rhea, and you may also spot Tethys and Dione, and perhaps Iapetus. Most people don't spot these moons because they are so excited by Saturn's rings, but they are worth looking for. You can find out which one is which online from www.calsky.com (go to Planets, Saturn, Apparent View/Data).

► Radar aboard Cassini reveals Titan's landscape to have lakes and rivers of methane or ethane. This is Ligeia Mare, a lake over 400 km across.

Uranus

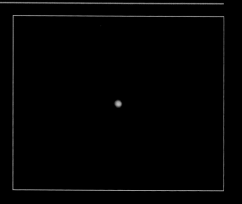

This was the first planet ever to be actually discovered, and it was first spotted from a back garden in an English town.

▶ Even a telescope as small as 125 mm as used for this photo can show the disk of Uranus, but don't expect to see a lot!

Where to look

The bright planets Mercury, Venus, Mars, Jupiter and Saturn were known about since prehistory, but it was not until 1781 that anyone knew the planet Uranus existed, because it is too faint to be obvious without a telescope. This means that you need to know its exact position on the night you'll be observing to be able to find it. The planet moves slowly through the sky, and is in the constellation of Pisces until 2018, and then in Aries until 2024. So it will be in the evening sky in the winter months.

Unless you have a Go To telescope which finds objects automatically, you will need a finder chart, which you can get online by searching for "finder chart for Uranus." Ideally you need two maps, at two different scales. The first will show you the major constellations, with just the brighter stars, so you can see which part of the sky to look in. The second shows you the area in detail where you can find Uranus and will show stars that are as faint or a bit fainter than the planet itself. Alternatively, download a free sky mapping program such as Stellarium that will let you zoom in on the planet and print out your own maps.

What you'll see

Though Uranus is not a star, it is so tiny in the sky that it is hard to tell the difference.

Uranus is about magnitude 6, so first check the sixth-magnitude stars in the area. The trick is to look for patterns of stars. Even from a town binoculars should show the planet, but if Uranus is not near any brighter stars you might have difficulty in finding the right spot in the sky. With a telescope, use the lowest magnification you have so as to get the widest field of view.

Fact File

Diameter	51,118 km
Typical distance from Sun	2,871 million km
Mass compared with Earth	14.5 times
Year length	84 Earth years
Day length	10h 6m
Typical cloud-top temperature	−195°C
Moons	At least 27

▶ Even the Hubble Space Telescope shows Uranus as a boring globe – so what chance do the rest of us have?

Uranus moves slowly through the sky, so you might have to wait a few nights before you can detect any shift in its position, proving that you really have found it using binoculars.

If you have a Go To telescope, you might think it will make the job easy. But Uranus looks so much like a star that it can be difficult to know that you've actually found it. Go To telescopes often do not put the object you want bang in the center of the field of view. At least in binoculars you can see nearby stars to help identify it, but a telescope shows a smaller part of the sky so you can't be so sure. In this case, you have to look at each of the stars in turn to see if it is any different from the others.

Once you think you have found Uranus, you can increase the magnification of your telescope to at least 150 to try and see its disk. This is not always obvious. Uranus's disk is only about 3½ arc seconds across. To show what that means, Jupiter is about 45 arc seconds across, so if Jupiter appears tiny in your telescope, Uranus will be really, really tiny. And you need good seeing as well so that the star images are not smeared out as much as the disk. The other thing to look for is the slightly bluish color of Uranus. This is not always as obvious as you might expect from artworks in books!

Not even large telescopes show real detail on Uranus, so don't expect to see anything much. The most they ever show is a slight darkening of one side. None of its moons are visible through small telescopes, either!

The discovery of Uranus

In March 1781, an amateur astronomer named William Herschel was using a home-made telescope from his backyard in Bath, England, looking for double stars. He came across one star in Gemini that looked quite different from all the others, with an actual disk rather than being a point of light. At first he thought he must have found a new sort of comet, particularly when four nights later he found that it had moved position. But eventually it became obvious that it was really a new planet, the first ever to be found rather than known about from ancient times.

▲ William Herschel discovered Uranus with a 150 mm home-built telescope with a metal mirror, probably this one.

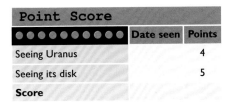

Point Score									Date seen	Points
● ● ● ● ● ● ● ● ●										
Seeing Uranus										4
Seeing its disk										5
Score										

The most distant known planet is a challenge for owners of small telescopes and binoculars.

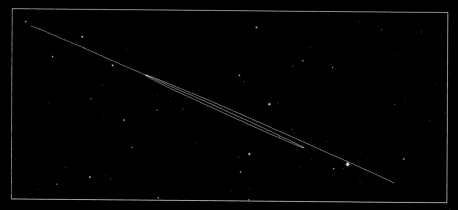

▲ The path of Neptune from January 2020 at far right until May 2021 at far left. It moves eastward until June 24, then starts moving westward until November 30 when it resumes moving eastward again – a retrograde loop. The bright star near the start of the track is fourth-magnitude Phi Aquarii.

Where to look

Neptune moves very slowly in its orbit, so it will be in the same part of the sky for years to come. It is in Aquarius until 2022, when it starts moving into Pisces. So it will be below the easily found constellation of Pegasus, and will be best seen when Pegasus is high up, which is in the autumn and winter months.

As with Uranus, but more so, you need to have maps that will show you exactly where to look, or use a Go To telescope. You can find finder charts on the web, or download a free sky mapping program such as Stellarium, or maybe you could use a tablet or phone app.

What you'll see

Neptune is fainter than Uranus but it is still within the reach of binoculars, though if your skies are really light polluted and the binoculars are quite small you will find it tricky. It looks just like a slightly bluish star, and not until you magnify it about 200 times with a telescope can you see that it actually has a disk.

Although Neptune is only slightly smaller than Uranus, and about four times the diameter of Earth, it is around 5 billion kilometers away so it isn't surprising that it's really tiny. The disk is just over 2 arc seconds across, which is nearly half that of Uranus. Quite often, Earth's atmosphere

Point Score		
●●●●●●●●●	Date seen	Points
Finding Neptune		5
Seeing its disk		6
Score		

smears images out more than this, and you'd need a telescope with an aperture of 100 mm or larger to see anything at all. Most people find that you really need about a 200 mm telescope to see the disk properly. And, like Uranus, no details are usually visible. So just be happy to see it at all!

Fact File

Diameter	49,528 km
Typical distance from Sun	4,495 million km
Mass compared with Earth	17 times
Year length	164 Earth years
Day length	16h 7m
Typical cloud-top temperature	−200°C
Moons	At least 14

▲ As with Uranus, there's little to see on Neptune other than a bluish disk, even using the Hubble Space Telescope.

Why doesn't Neptune move smoothly?

When you look at a map of the track of Neptune over the year, you'll see that it doesn't just march steadily eastward in its orbit as you'd expect. In June it comes to a standstill, then starts moving backward until November or December, then carries on moving westward again. What's happening is that the Earth is moving faster in its own orbit around the Sun, so we are overtaking it on the inside track and for a while it seems to move backward.

This happens with all the planets that are farther than us from the Sun. It is called *retrograde motion*, and when a planet is moving the right way it is called *prograde motion*.

23

Asteroids and dwarf planets

These bodies orbit the Sun, like planets, but are much smaller – and much harder to see.

▲ These two pictures of Vesta were taken two days apart. See if you can pick out which one it is.

Where to look

The asteroids, or minor planets, orbit the Sun mostly between the orbits of Mars and Jupiter. As they are much smaller than the main planets, and are quite distant, you have to know exactly where to look for them. As with the fainter planets Uranus and Neptune, this usually means finding their positions on a website or using an app. But there are only a few minor planets bright enough to be easily visible using small telescopes and binoculars.

The website www.heavens-above.com lists those asteroids that are brighter than magnitude 10, which will be visible with small telescopes, and gives you two charts.

Point Score		
●●●●●●●●●	**Date seen**	**Points**
Finding an asteroid		6
Seeing that it has moved on a separate night		3
Score		

One shows the area of sky where the asteroid is at the moment, so you can work out whether it will be visible in the evening sky by checking with the charts in this book or on a sky map app. Try to choose a time when the asteroid is high in the sky, and more or less in the south.

The other shows just the area of sky around the asteroid. This covers 2° of sky, which is not as much as you see with ordinary binoculars (usually about 5°), so it can be tricky working out which star is which.

The free download star atlas Stellarium also includes some asteroids. Search for the brightest ones, which are Ceres, Vesta, Pallas and Juno. See if any are around in the evening sky at a time when you can observe using our star maps on pages 218 to 223.

Apps for phones and tablets may also allow you to search for any asteroid, even very obscure ones, and will display them on their maps, but it's best to stick to the brightest to start with. As well as

When is a minor planet a dwarf planet?

Back in earlier times, everyone knew what a planet was – it orbited the Sun and was pretty big. Then in the 19th century people started to discover lots of much smaller objects between Mars and Jupiter, that didn't really qualify as planets though they orbited the Sun, so they called them minor planets or asteroids.

In 1930, Pluto was found beyond Neptune, and while it was pretty faint, it was the only one of its type and was probably about the size of Mercury, so people were happy to call it a planet. But then in the 1990s new bodies about the size of Pluto started to turn up at the edge of the Solar System, and it became clear that Pluto was not as big as people thought. So were these new objects planets as well?

In 2006 it was decided to have a new category: dwarf planet. This was bigger than a minor planet, but was not big enough for its gravity to prevent other bodies from orbiting in its own region of space. Ceres turned out to be big enough, because it is a sphere instead of being an irregular shape like Vesta. However, it has to coexist with minor planets such as Vesta, which are roughly the same distance from the Sun, so it can't be called a planet.

Poor old Pluto was demoted from being a planet to a dwarf planet, as it was not big enough to prevent other similar bodies from orbiting near it. But some astronomers disagree with the definitions, so things may change yet again.

Because asteroids and dwarf planets are small, very little is known about them compared with the major planets. Most of them have not been photographed in detail.

the six listed below, the next brightest are Iris, Flora, Metis, Irene, Massalia and Bamberga. These are not actually the biggest, but they have brighter surface material than some larger asteroids.

What you'll see

An asteroid appears in binoculars or a small telescope as a point of light, just like a star, so it won't exactly jump out at you when you see it. The only way to be sure that you've seen it is to make a note of its position

▲ Ceres (left) and Pluto, to the same scale of their diameters. Both were taken using the Hubble Space Telescope, but Pluto is so far away that the details are very vague.

compared with nearby stars and look again either the next night or whenever it is clear. The asteroid will have moved.

Fact File

Name	Diameter	Typical distance from Sun	Year length
Ceres	975 km	414 million km	4.6 Earth years
Pallas	544 km	415 million km	4.62 Earth years
Vesta	525 km	353 million km	3.63 Earth years
Hygiea	431 km	470 million km	5.56 Earth years
Juno	258 km	400 million km	4.37 Earth years
Pluto	2,300 km	5,900 million km	248 Earth years

Observing tip

Asteroids are not always easy to spot, and often the problem is finding the right bit of sky to start with. So if you have an online star map or an app, look for a date when one is close to a brighter star that you can easily find. This can avoid a lot of bother!

24

Comets

Everyone wants to see a comet – but bright ones are rare, and finding the more common fainter ones can be a challenge.

▶ Most comets are faint and can only be seen with binoculars or telescopes, such as Comet Garradd 2009 PI, which was seen in 2012. The star images trailed during the exposure due to the movement of the comet through the sky.

Where to look

Comets are city-sized chunks of ice from the outer Solar System that get diverted into orbits that bring them close to the Sun. They don't usually have almost circular paths like the planets, but have long looping orbits that take them out into the far reaches beyond Jupiter, then back in toward the Sun where they spend a few weeks or months. There are far more comets out there than we know about, so new ones can appear at any time and come from any direction. This makes it impossible to give general hints about where to look – you have to check websites that list currently observable comets.

The most useful site for beginners is www.heavens-above.com. Here, you'll see a listing of whatever comets are currently visible in medium-sized telescopes, with star maps of their positions. But you'll need to work out for yourself whether the comet is bright enough to be seen using your telescope, and whether it is in a part of the sky that you can view. Bear in mind that it might be close to the Sun, so even if it's above the horizon you should not look for it while the Sun is in the sky. See the Tips section opposite.

Heavens-above.com gives two maps, one showing the area of sky where the comet lies, and another showing a detailed view of the sky in the small square in the center of the map, which covers 2° of sky. This is quite a small area, so if the comet is faint you will need to use a lot of skill in finding the right spot.

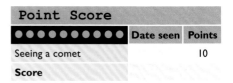

Point Score	Date seen	Points
●●●●●●●●●●		
Seeing a comet		10
Score		

What you'll see

Pictures of comets often look as if they are shooting rapidly through the sky, with a tail streaming behind. But they are way out in the Solar System, not close to Earth, so they move only very slowly through the sky and they rise and set along with the stars. You only see the movement from night to night, or sometimes by watching carefully using a telescope or binoculars.

There is a central bright area, or nucleus, surrounded by a haze, or coma. In some comets there is a tail, or sometimes more than one tail, consisting of material that has come from the nucleus and is starting its own orbit around the Sun. The size of the comet as seen in the sky can vary a great deal – some are large enough to be easily visible with binoculars, but others require telescopes and a lot of magnification.

Brain Box

Are comets poisonous? They often contain a molecule of carbon and nitrogen – cyanide! But even if Earth passed right through a comet's tail, the amount of it would not harm us.

Tips for viewing comets

Once you know which constellation the comet is in, use the star maps on pages 218–223 to find when that part of the sky is visible. It may not be easy to observe in a twilight sky, or it may be only visible from the southern hemisphere.

Comets often appear fainter than their magnitude brightness suggests. For example, if a comet is supposed to be magnitude 6, you would think it should be easily visible in binoculars, but if it is very spread out and hazy, you may find that it isn't visible at all. Or it might be that you have to use a telescope, which has greater magnification than binoculars. If the comet is in the twilight sky, or there is lots of light pollution around, it might also be the same brightness as the background and impossible to see. You need a lot of patience with comets.

▶ Occasionally a bright comet appears, such as Comet Hale-Bopp, which was easily visible with the naked eye for several weeks in 1997 even from suburbs and city centers.

Fact File

Typical comet diameter	10 km
Typical comet orbital period around the Sun	Thousands of years
Typical comet composition	Water ice, frozen carbon monoxide, carbon dioxide, methane, dust

What was that? Suddenly a star seems to flash across the sky. Many people call them shooting stars, and astronomers call them meteors.

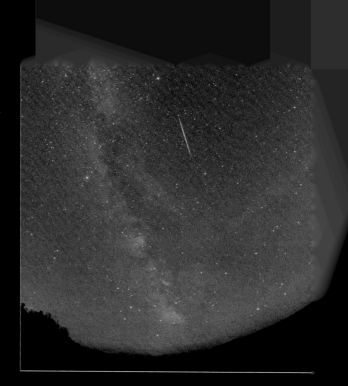

▶ A meteor dashes through the sky. Photographing meteors is a matter of luck because you can never tell when one will suddenly appear.

Where to look

A meteor can happen at literally any moment, anywhere in the sky, without warning. But there are some nights when you have a better chance of seeing one, and a "meteor shower" is expected. These occur around the same dates each year, when Earth plows through a denser stream of the tiny particles that create meteors.

These bodies are called meteoroids, and they are really the dust from the tails of comets that crossed this region of space possibly hundreds or thousands of years ago. They collide with Earth's atmosphere at a speed of many kilometers a second, and friction with the air molecules burns them up and makes the air glow briefly.

Each particle is only tiny, and it vaporizes about 80 km up. There's no danger that one of them will fall to the ground so it's perfectly safe to go out and watch for them. The only way to see one is just to gaze at the sky and wait for one to appear.

The table opposite lists the dates on which meteor showers are active. The name of the shower tells you which constellation the meteors will appear to radiate from, though you could see them

Point Score

●●●●●●●●●	Date seen	Points
Seeing a meteor		2
Carrying out a meteor watch and making notes		3
Score		

Dates of meteor showers			
Date	Shower name	Radiant	Rate
Jan 3–4	Quadrantids	Boötes	High
April 22	Lyrids	Lyra	Low
May 5–6	Eta Aquarids	Aquarius	Medium
July 29	Delta Aquarids (1)	Aquarius	Medium
Aug 6	Delta Aquarids (2)	Aquarius	Low
Aug 12	Perseids	Perseus	High
Oct 21–24	Orionids	Orion	Medium
Nov 5	Taurids (1)	Taurus	Low
Nov 12	Taurids (2)	Taurus	Low
Nov 17	Leonids	Leo	Low to medium
Dec 14	Geminids	Gemini	High
Dec 22–23	Ursids	Ursa Major	Low

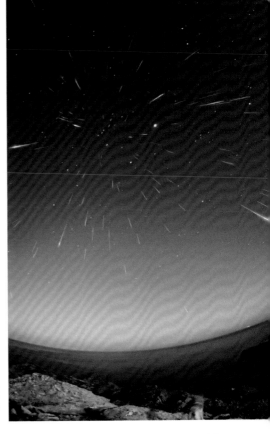

▼ During a meteor storm many hundreds of meteors can appear in an hour. This is a combined exposure of the Leonids in 2002.

in any part of the sky. Although they are on parallel paths as they travel through space, when they appear in the sky the tracks appear to come from a single point, called the radiant.

Although these are the best dates, meteors are more common between August and December than from January to June. There can also be meteors which are not linked to any particular shower, called sporadic meteors, caused by dust grains that have spread a long way from the orbit of their comet over thousands of years.

Very rarely there are meteor storms, where hundreds or even thousands of meteors can appear per hour. This happens only a few times each century, and they last perhaps for a few hours

Fact File

Typical meteor height	80 km
Typical meteoroid mass	A millionth of a gram to 1 gram
Typical meteoroid velocity	18 to 70 km/s

▼ A typical small meteorite has a black exterior caused by heating during its fiery trip through the atmosphere. But not all black rocks are meteorites!

What's a meteorite?

People sometimes call shooting stars meteorites, but this is not correct. A meteorite is a chunk of material from space that has fallen to Earth. You can remember the difference because the names of geological specimens often end in "ite," such as "stalagmite" or "dolomite." Meteorites are pieces of asteroid, and can very occasionally be large enough to survive their fall through the atmosphere.

so the high numbers might appear at a time when it is daylight or cloudy where you are.

What you'll see

A meteor usually lasts only a fraction of a second, so there's no hope of looking at it even through binoculars, let alone a telescope. It dashes through a part of the sky, usually only covering about a single constellation. There are bright ones and faint ones, just like stars, and the range of brightness is about the same as the stars, though it is quite tricky to see the very faint ones.

Some meteors leave a train behind them, which usually lasts for a fraction of a second but sometimes can linger for seconds or even minutes.

Meteors don't always move downward. They can travel upward, particularly when you see them close to the horizon, and can sometimes appear to come straight toward

Brain Box

Meteors brighter than magnitude −5 (brighter than Venus) are known as fireballs. Often these are not ordinary meteors but are caused by larger bodies that may fall to ground as a meteorite.

you. On a good night you might see one meteor every few minutes, maybe with short bursts of two or three a minute. But even on shower nights you can go for ages without seeing a single one.

If you see something which moves quite slowly, and lasts longer than a second, it could be a **satellite** or an **Iridium flare** – that is, as long as it isn't a plane!

Observing tips

Lie on a lounger or deckchair, and keep a blanket handy even on a summer's evening. Try to avoid any direct lighting, and keep watch on as much of the sky as you can. When a meteor appears, make a note of the time, and compare its magnitude with stars whose brightness you know. Alternatively, just count the number that you see in, say, half an hour. It's best to watch for at least an hour to give yourself a good chance of seeing some.

These mysterious and beautiful night clouds are quite unlike ordinary ones and are actually on the edge of space.

◀ Sometimes noctilucent clouds can fill the whole northern sky. Ordinary clouds appear dark against the display.

Where to look

These clouds are only around on clear nights in midsummer, from about mid May to mid July, and usually appear only very late at night, when the rest of the sky is dark. They are seen only on the northern horizon, and are usually visible only from more northerly latitudes, above about latitude 50°, but sometimes from the northern US. They don't appear every night, so you have to be lucky as well.

What you'll see

Noctilucent clouds (often just called NLCs) look very like ordinary cirrus clouds, but they are visible on some nights along the northern horizon well after the Sun has set in midsummer. So to see them you would have to be up late, after 11 pm, and with a clear view of the northern horizon.

Have you seen NLCs or are they cirrus clouds? Cirrus just after sunset can look a bit like NLCs, but they don't appear as late in the evening. Another clue is that you can sometimes see ordinary clouds silhouetted against the NLCs, showing that the NLCs are much higher.

There seem to be more NLCs now than in the past, and such things as climate change and pollution from aircraft engines may have something to do with this.

Fact File

Name	Noctilucent is Latin for "night-shining"
Typical height	80 km
Composition	Ice crystals

Point Score

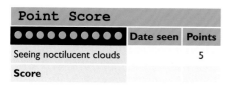

●●●●●●●●●	Date seen	Points
Seeing noctilucent clouds		5
Score		

27

Northern Lights

People travel to the Arctic Circle to view this amazing spectacle – but you could be lucky enough to see it from your own backyard.

▶ A major aurora seen from Oxfordshire, UK, in 2000. These curtain-like forms are more usually seen much farther north.

Where to look

Although the Northern Lights, or the aurora as astronomers say, can be seen on many nights of the year if you live in northern Alaska or Canada, the rest of us have to get lucky to see them. But they do appear farther south, and really strong displays can very occasionally be seen as far south as southern California or Florida. However, the farther north you are, the better.

The Northern Lights are caused by storms of particles from the Sun, so they are more common when the Sun is active and has lots of spots on it, but there's no guarantee that a spotty Sun means that you'll see an aurora that night. There has to be a flare on the Sun at just the right time sending a stream of particles toward the Earth.

But when the Sun is active (see page 95 for clues) it is worth keeping an eye on the northern horizon after dark in case there is an aurora. Get to know the appearance of the sky in that direction so if there is an aurora you can tell it apart from light pollution.

Faint displays can actually be seen from northern US states and similar latitudes fairly often when the Sun is active, though they don't make the headlines. There are websites (such as aurorawatch.ca and spaceweather. com) where you can keep track of the Sun's activity and discover the chance of there being an aurora at any time.

Fact File

Name	Aurora was the Roman goddess of the dawn. The Northern Lights are known as the *aurora borealis* and similar lights seen in the southern hemisphere are the *aurora australis*.
Typical height	80 km
Composition	Green is due to glowing oxygen atoms in the upper atmosphere; red is due to glowing nitrogen atoms.

What you'll see

The Northern Lights usually appear as red or green bands of light, though some other colors are possible. The color is not always obvious, but shows up in bright displays. Often a display starts with a horizontal green or red glow, covering quite a large area of the northern sky. This can develop vertical bands or rays, and in the really super displays these shift around as you watch. The movement is not usually rapid, but it's quick enough that you can see changes every few seconds.

Very occasionally a strong display can cover the whole sky, and streamers seem to be coming from directly overhead. These usually make the news, but usually only after they have happened and you've missed them because you were indoors!

Observing tips

If you aren't sure whether you are seeing the Northern Lights or just plain old light pollution, the trick is to take a time exposure on a digital camera. Most cameras can give exposure times of 15 seconds or so, and at sensitivities of ISO 1600, which is what you need. So read the instructions and work

Point Score		
●●●●●●●●●●	Date seen	Points
Seeing an aurora		5
Seeing a really bright aurora without cheating and going on an expensive Northern Lights trip		10
Photographing an aurora		5
Score		

out how to switch off the flash and give time exposures at ISO 1600 with the focus on infinity (usually shown as a mountain symbol on the scene guide). Some phone cameras will do this as well. A time exposure is different from delayed shutter, which takes an ordinary picture after 10 seconds to allow you time to get into the picture yourself.

Hold the camera firmly against a solid object pointing at the sky on a wide-angle lens setting and press the shutter. If there is an aurora present, you will see green or red bands. If all you see is a boring orange glow, no luck. But you should have photographed some stars as well, and it's a good idea to experiment with doing this on a clear night anyway so you get used to doing it.

Soundbite

Aurora is pronounced "or-ora." The plural of aurora is aurorae (pronounced "or-or-ee") but "auroras" is used in books that don't believe we should use Latin plurals!

▶ It wasn't clear at first whether this glow behind clouds was an aurora. But a 30-second exposure on a compact digital camera showed the red color.

28

International Space Station

It's sometimes the brightest object in the night sky after the Moon, and the ISS is a great object to show off to your friends.

◀ The ISS appears just as a dot of light, but a time exposure shows a trail as it crosses the sky. It fades as it enters Earth's shadow.

Where to look

The International Space Station orbits Earth roughly every 90 minutes, but its track is different on every pass. It is easily visible from the ground when passing over, as long as it happens to be in sunlight as it does so. This applies before sunrise and after sunset at all times of year, and between April and September you could see it at most times of the night as well, as the Sun is not too far below the horizon so the Station is still sunlit.

But there is a lot of the Earth for it to pass over, and on some orbits it is going to be too far south for you to be able to see it, or even over the southern hemisphere. So you have to be watching at exactly the right time. How do you know when? You need to find out online, either at www.heavens-above.com or at spaceflight.nasa.gov/realdata/sightings/.

In each case, you first have to enter your location, and then you will get a table of sighting times which give you the direction where it is first visible and the height in degrees, its maximum altitude and the direction where it is last visible for each sighting. On heavens.above.com there's also a sky map showing the track, which will help you to look out for it. It always moves from west to east, but don't expect it to rise due west and set due east – it's usually one side or other of those directions. Sometimes it can be low down and sometimes high up.

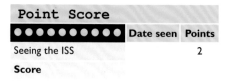

Point Score		
●●●●●●●●●●● Date seen	Points
Seeing the ISS		2
Score		

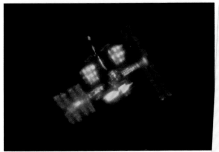

Fact File

Typical height above Earth's surface	420 km
Orbital time	92 minutes
Orbital speed	28,000 km/h or 7.7 km/s
Dimensions	73 × 109 × 20 m
Mass	450 tonnes
Typical magnitude when overhead	−3.5

▲ This close-up of the ISS as seen from the ground was made by combining many video shots taken through a 250 mm telescope.

What you'll see

The ISS looks just like a very bright white star moving steadily through the sky. It moves faster than most aircraft unless they are quite close to you, in which case you'd see the wing lights as well and probably hear them. It always moves at the same speed, crossing the sky in about 7 minutes at most, and moves in a straight line.

Quite often it fades out, which is because it has moved into the Earth's shadow, and the Sun sets for those on board. If you are viewing it in the early morning, it could brighten up instead as dawn breaks over the station. We see the ISS only when it is in sunlight – any lights aboard it are not nearly bright enough to be seen from the ground.

As the ISS moves quite quickly through the sky, it is rather hard to follow using a telescope. But if you do catch a glimpse of it, using a magnification of about 50 or more, you might just be able to see its H-shape. Occasionally you might see another, fainter object following it. This will be a supply vessel, such as a Soyuz manned craft or a Progress cargo vessel.

The ISS usually carries six people. A tour of duty aboard lasts up to six months. They travel to and from the station aboard Soyuz craft, each of which can carry three people. There are always sufficient craft available to evacuate everyone in case of an emergency.

The station has giant sets of solar panels to generate electricity, and it has separate modules for the crew's living quarters, service areas to keep the station going, and various laboratories. One thing it lacks is a dormitory module. Instead, there are one-person crew cabins around the station, with sleeping bags. No need for a bed when you are weightless!

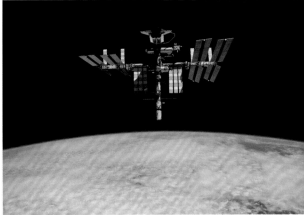

▶ The International Space Station as it appears from an approaching spaceship.

You can amaze your friends with your powers of prediction using these bright and spectacular satellite glints.

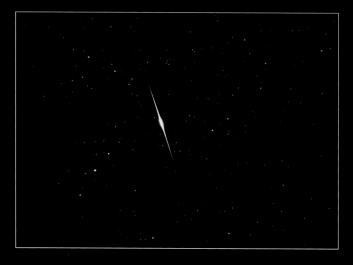

◀ An Iridium flare close to the star Altair in Aquila. The event takes about 10 seconds to occur.

Where to look

An Iridium flare is a reflection from the polished surface of a satellite in the Iridium series, which orbit the Earth every 100 minutes. So to observe one, you need an accurate prediction of where and when it will appear in the sky. There are many places where you can find these predictions, the most popular being phone apps and the website www.heavens-above.com.

You'll need to enter your location (within a few kilometers), which in the case of a phone app may use the phone's GPS receiver. On heavens-above.com, once you have registered and logged in (free) you can choose your location from a large database. The predictions give time of event, brightness (in magnitudes – see page 16), altitude (in degrees above the horizon) and azimuth (in degrees from north or compass direction). There's also a sky map, which shows you exactly where you should be looking at the right time.

Armed with this information and an accurate watch, you can amaze people by suddenly pointing at the sky and conjuring up out of nothing what looks like a very bright moving star! But the problems come when they want another one a few minutes later and nothing is predicted....

Point Score		
●●●●●●●●●●	Date seen	Points
Seeing a predicted Iridium flare		2
Amazing your friends		3
Photographing a flare		3
Score		

Tip: practice first on a night when there's no one else around! For best results choose a bright one, which means with a magnitude of at least –2, remembering that minus figures are the brightest.

What you'll see

If you've ever seen a **satellite** or the **ISS** you'll know that these usually move steadily across the sky and remain at more or less the same brightness unless they go into Earth's shadow. Iridium satellites move in the same way, a bit slower than the ISS, but because their polished panels can catch the sunlight they suddenly flare up in brightness by many times. So to begin with they are just ordinary satellites, probably too faint to be seen, then over a matter of a few seconds they flare up and then just as quickly fade away again until you can no longer see them.

You can photograph the event by pointing a camera at the right part of the sky, firmly fixed in position, and giving a time exposure which will just cover the appearance of the flare. Many cameras these days will give a 15-second exposure, which is just long enough (see page 83 for more tips on sky photography). In a dark sky use ISO 1600, but if it's light, try 400 or even 100. Experiment in advance!

Fact File

Purpose	Used for satellite phones
Number	66 satellites plus spares
Orbital period	About 100 minutes
Orbital height	About 780 km

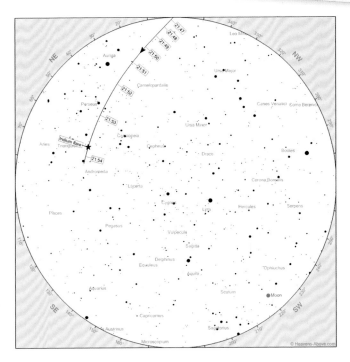

▶ This screen shot of heavens-above.com shows the track of the satellite and the center of the flare, which is the spot you need to watch.

Earth satellites

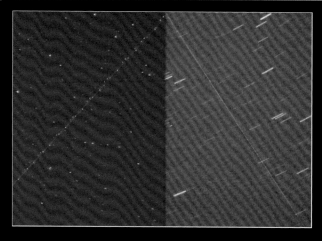

These things are orbiting the Earth all the time, and they are surprisingly easy to spot, particularly in summer.

◄ On the left, an aircraft passes though a photo of star trails, with flashing lights leaving a repeating pattern. On the right – the real thing, a satellite trail.

Where to look

You might see something moving steadily through the sky that doesn't seem to be a plane, just by chance. There are thousands of artificial satellites in orbit, and hundreds that you might see. But the best thing is to be able to predict when one is due. You can do this using www.heavens-above.com, or you may find a phone app that will do it.

The information given by heavens-above.com is the same as for predicting **Iridium flares** or the **ISS**, so use the information given there to log on and predict a satellite. Most satellites predicted are around magnitude 2 or fainter (which means larger magnitude numbers), so to start with choose the brighter ones. The map will show its track through the sky, but bear in mind that sometimes it may only be visible for a short time close to the horizon so it may not be obvious.

What you'll see

A satellite usually looks like a star moving steadily through the sky – just a point of light. If you can see other lights with it, it's probably a very high plane! Its speed depends on how high it is. Some move only slowly, but others move as fast as the ISS. Occasionally the satellite may vary in brightness, if it is tumbling end over end. This applies especially to rockets, which are just the spent casings used to get the satellite into orbit. Sometimes all you see is just the occasional flash, which repeats as the satellite moves across the sky.

Like the ISS, the satellite is only visible if it is in sunlight, so it may fade in or fade out as it leaves or goes into Earth's shadow.

Point Score		
●●●●●●●●●●	Date seen	Points
Seeing a satellite by chance		2
Seeing a predicted satellite		2
Score		

From time to time the Moon passes through the Earth's shadow, so its surface goes dark – an eclipse of the Moon, or lunar eclipse.

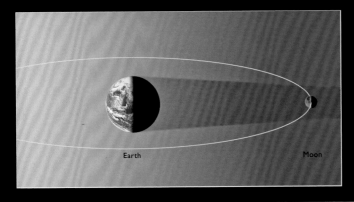

Earth

Moon

◄ A lunar eclipse happens when the Moon passes through the Earth's shadow. This diagram isn't to scale – the Moon is really much farther away from the Earth.

When to look

Eclipses can only happen at certain times when the Sun, Earth and Moon are exactly in line (what's called *syzygy* – a good word to have up your sleeve). You might think that this would happen every month at full Moon, but because the Moon's orbit around the Earth is at an angle to the Earth's orbit around the Sun, it only happens every 25 weeks or so. Then, there is the chance of both lunar and **solar eclipses** taking place within a few weeks.

A lunar eclipse can be seen from anywhere that the full Moon is visible. But whether or not you see it depends on where you live, because it might be happening on the other side of the world. The list on page 91 shows all the lunar eclipses that will be worth observing from the US until March 2026, with the time of mid-eclipse. There are others, but you would have to travel to another part of the world to see them.

In some cases, only a small part of the whole eclipse might be visible from your location. Unfortunately, there are no total eclipses well visible in the evening until 2028 and 2029.

What you'll see

As the Moon starts to go into the Earth's shadow, its eastern side begins to get dimmer. The shadow doesn't have a sharp edge, so you can hardly notice the difference at first. The outer part of the shadow is called the penumbra. In some eclipses, the penumbral ones, the Moon only goes through the penumbra so they are not very spectacular.

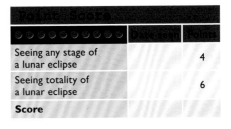

Point Score		
	Date seen	Points
Seeing any stage of a lunar eclipse		4
Seeing totality of a lunar eclipse		6
Score		

If the Moon goes more centrally through the shadow, it goes into the umbra, which is much darker. At this point, people who aren't familiar with looking at the Moon may say that it often looks like that – but they are confusing the phase of the Moon with the Earth's shadow. When you look at it through binoculars or a telescope you can see that there is no terminator and none of the craters are side-lit, as happens during the usual phases. But the edge of the shadow is curved, because the edge of the Earth's shadow is circular.

Sometimes the Moon only goes into part of the umbra, so only the north or south part of the Moon is in shadow. This is a partial lunar eclipse. But if you're lucky, the Moon goes right into the shadow, and you see a total lunar eclipse. The whole Moon goes dark.

Even so, it doesn't necessarily disappear from the sky. Instead, it usually goes a reddish color. This is because some of the Sun's light is beamed around the edge of the Earth, and reddened by the atmosphere in just the same way that the Sun is reddened when we see it at sunset. If you were on the Moon during a total lunar eclipse, you'd see the Earth up there in the sky as a red ring, with the Sun hidden behind it.

Brain Box

Ancient peoples are said to have feared that the Moon was being eaten during an eclipse, so they made lots of noise to scare the monster away. How fortunate that they did, as the Moon is still with us!

Fact File

Stages of a partial lunar eclipse

First contact	Moon starts to enter penumbra (outer shadow).
Second contact	Moon enters umbra and partial eclipse begins.
Third contact	It's fully within the umbra and totality begins.
Greatest eclipse	The Moon is in the darkest part of the shadow.
Fourth contact	It starts to move out of the full shadow.
Fifth contact	The Moon leaves the umbra.
Sixth contact	Moon leaves the penumbra and it's all over.

Some eclipses are darker than others. During a lunar eclipse the stars come out, and during a dark one if you didn't know better you wouldn't even notice the Moon. Just imagine what our ancestors must have thought when the Moon they relied on for finding their way at night had faded away. Anyone who could predict eclipses would have been on to a good thing.

Lunar eclipses visible from the US (given in Central Time, not allowing for Daylight Saving)			
Date	Mid-eclipse (CT)	Type	Comments
2015 April 4	06:00	Total	One for early risers.
2015 September 27	20:47	Total	Excellent evening eclipse.
2016 March 23	05:47	Penumbral	Occurs at moonset.
2016 August 18	03:42	Penumbral	Seriously unspectacular.
2017 February 10	18:44	Penumbral	Rises during the eclipse.
2018 January 31	07:30	Total	Sets during the eclipse.
2019 January 20	23:12	Total	High in the sky at totality.
2020 July 4	22:30	Penumbral	Watch it during the fireworks.
2020 November 30	03:43	Penumbral	Only for the dedicated.
2021 May 26	05:18	Total	Moon sets eclipsed.
2021 November 19	03:03	Partial	Almost total.
2022 May 15	22:11	Total	Good evening eclipse.
2022 November 8	04:59	Total	Ends after sunrise.
2024 March 25	01:12	Penumbral	Almost a partial.
2024 September 17	20:44	Partial	Only a small partial.
2025 March 14	00:58	Total	Stay up late.
2026 March 3	05:33	Total	Another early start.

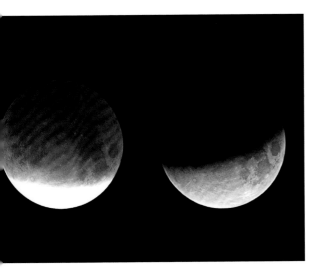

◄ Sequence of events during a lunar eclipse. The Moon moves from right to left through the Earth's shadow, so the right-hand shot is the first in the sequence.

91

Even though we rely on the Sun so much, many people don't really understand how it moves through the day and through the year.

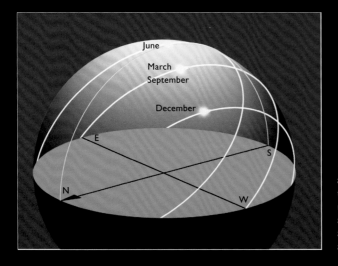

June
March
September
December

E
S
N
W

◄ The Sun's movements are easiest to follow if you think of the Sun as going round the Earth. Only at the equinoxes does it actually rise in the east and set in the west.

Where to find it

This isn't as obvious as it seems. Yes, on a clear day you can find the Sun without too much trouble, but as astronomers we need to know a bit more about its movements. Everyone will tell you that it rises in the east and sets in the west, but most of the time this isn't true! Only twice a year does it do this, and the rest of the time it rises and sets to one side or the other of these points.

In winter, it rises over in the southeast, climbs a short way in the sky and then sets in the southwest after a few hours. In summer, it rises in the northeast and gets very high in the sky at noon, finally setting after many hours in the northwest. The midwinter and midsummer positions are called the *solstices*. In spring and autumn it is halfway between these points, and only then (on March 20 and around September 22) does it actually rise due east and set due west. These times are called the *equinoxes*.

What you'll see

First, and this really is important, *never* look directly at the Sun, even without a telescope. OK, I know people often do what they are not supposed to just to prove that they can't be told what to do, but there's no advantage in ruining

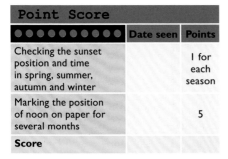

Point Score		
⚪⚪⚪⚪⚪⚪⚪⚪⚪⚪	Date seen	Points
Checking the sunset position and time in spring, summer, autumn and winter		1 for each season
Marking the position of noon on paper for several months		5
Score		

your eyesight just to make a point. The Sun really can blind you.

When you do catch a glimpse of the Sun, such as at sunset, you don't actually stare at it, and your instinct is to look away, so it does little harm. To actually study the Sun means gazing at it for a long time, so you need special equipment, as explained in the article on **sunspots**. Here we are dealing with the Sun's movements through the sky.

From your home, make a note of where on the horizon the Sun sets every so often, and the time. In winter, do this in the morning as well, but eventually you'll find that the Sun is well up by the time you are up and about. Also keep a note of the Sun's position at midday. The best way to do this is to set up a vertical stick and mark the position of the tip of its shadow on a sheet of paper as close as possible to noon every week or so whenever you can. This

could be in a sunny south-facing window or outside, but you need to be able to put it all back in exactly the same place each time you make your measurement.

When the clocks go forward or back, you'll notice that there is a sudden change in the shadow position at noon. This is because in summer, midday is not around 12 o'clock, but about 1 pm as the clocks have changed.

The exact moment when the Sun is highest in your sky, and is exactly south, depends on a number of things. It varies slightly throughout the year because of variations in the speed of Earth's orbit around the Sun, and there is also a fixed difference that depends how far east or west you live of your time meridian. Eastern Time, for example, is based on longitude 75°W (Philadelphia). For every 1° of longitude difference, noon occurs 4 minutes later if you are west, or earlier if you are east. So in Indianapolis, which is at longitude 86°W, noon is 44 minutes later than in Philadelphia.

Fact File

Diameter	1,394,000 km
Typical distance from Earth	149,600,000 km
Temperature	5,500°C

▶ In midwinter the Sun is low in the sky at midday, but in midsummer it is at its highest, as these two shots from the same place show.

Sunspots – dark areas on the Sun – are a sign of solar activity. They vary in numbers over the years, and can even affect Earth.

◄ Projecting the Sun's image using a telescope (here, a reflector) doesn't require expensive apparatus. Any piece of white paper will do, and a box lid helps to keep the direct sunlight off the image.

Where (and how) to look

The crucial thing is that you should *never* look directly at the Sun with your unaided eyes, binoculars or a telescope. It is so bright that even a glimpse can cause blindness. But there are ways of observing sunspots using binoculars or a telescope that are quite safe, by projecting the image or using special filters.

To use binoculars to project the Sun's image, first cover one side of them to prevent accidents. Next, hold a piece of card about 30 cm behind the binoculars and point them at the Sun. You'll see a bright spot on the card, which is an image of the Sun. You can focus this image to get a clear circular disk, maybe 1 cm across. To make the image larger, hold the card farther away from the binoculars and refocus. The image will be larger but dimmer.

The procedure for a telescope is the same, but you must make sure that if it has a small finder telescope on top of it, this is also covered to prevent anyone from taking a peek when you are not expecting it.

Many telescopes have a special aperture in the main lens cover to restrict them to about 50 mm aperture on grounds of safety. Don't leave either binoculars or telescope pointing at the Sun for more than a few seconds at a time, and don't let the Sun drift out of the field of view. If there are plastic parts in the eyepieces its heat could burn a hole in the eyepiece body.

The other method is to use a special filter bought for the purpose of solar observing. This fits over the top of the telescope, so that only filtered light enters it, so you can then use a standard eyepiece to look directly at the dimmed view.

Fact File

Sunspot maximum	2024 ? (uncertain)
Sunspot minimum	2019 ? (uncertain)
Temperature	3,000–4,000°C

You can get sheets of suitable material, such as Baader Astro-Solar, to make your own filters but you must take great care that your filter can't become dislodged while you are observing. Never use a substitute that is not designed for solar observing, even if it looks dark, because it may let through harmful infrared rays.

What you'll see

Any sunspots will be visible as dark spots on the bright disk of the Sun. If someone tries to tell you that you are just viewing dust on the lens, try turning the eyepiece or the whole telescope or binoculars – the sunspots will remain stationary on the disk.

Sunspots move across the disk over a period of days, changing from day to day as they go. You might notice that they are in pairs, all aligned roughly in the same direction which is parallel to the Sun's equator. Make a drawing each day and you can plot the movements and changes.

You might also notice that the limb (edge) of the Sun is slightly darker than the center. This is called limb darkening, and is caused by the Sun's atmosphere. Sometimes you can see brighter areas known as faculae in this zone. Both sunspots and faculae are regions where magnetic fields affect the output of light from the Sun. Even sunspots are still quite bright, but they just look dark against the bright solar disk. A large sunspot can easily be much larger in diameter than the Earth.

Sunspots vary in size and number in a cycle of about 11 years. Sometimes whole months can go by without any being visible, while at other times they are plentiful. These times are known as solar minimum and solar maximum, and when the Sun is at maximum the **Northern Lights** are much more common. But the dates of maximum and minimum are very hard to predict, and they also vary in strength. The maximum of 2013–14 was particularly weak, and left people wondering if the Sun is becoming unusually quiet.

The largest sunspots can be 80,000 km in diameter – that is, about ten times the diameter of the Earth.

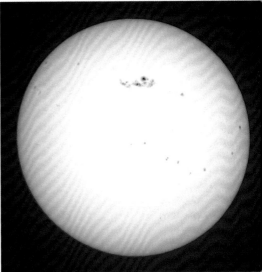

▲ The Sun on a spotty day. But don't be surprised if you see fewer than this. Sometimes the Sun can be spot-free for months.

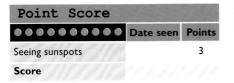

Point Score	Date seen	Points
⬤◯⬤◯⬤◯⬤◯⬤◯		
Seeing sunspots		3
Score		

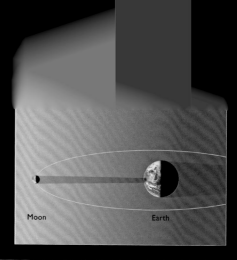

It's one of the most spectacular sights in the sky – but you will probably have to travel a long way to see this dramatic event.

▶ When the Moon gets in front of the Sun we see a solar eclipse. If you are on the edge of the shadow spot you see a partial eclipse, but to see a total eclipse you need to be near the middle.

Moon
Earth

When to look

Solar eclipses occur when the Sun, Moon and Earth are in line and the Moon's shadow crosses Earth's surface, which happens every six months or so. The Moon's shadow is over 3,000 km across, but you have to be right in the middle of it to see a total solar eclipse, where the Moon totally blots out the Sun. This is a strip only a couple of hundred kilometers wide.

The list opposite shows where you'll have to go to see a total eclipse over the next few years. Those marked (A) are annular eclipses. The trick is to find the place with the best prospects of clear skies. Many "eclipse chasers" treat these events as a good excuse to travel to far-off lands that they might not otherwise visit.

What you'll see

You really have to be careful when observing the Sun. It's so bright that even a sudden glimpse of it can cause eye damage or blindness. Even if the Moon is almost covering the Sun, the remaining

bit is just as bright and can still blind you. Read the **sunspots** article to find out how to observe the Sun safely. Alternatively, shops that sell telescopes may also have special eclipse viewers in stock, so search on the web for these well before the eclipse. They have small filters the size of spectacle lenses which are only suitable for looking at the Sun directly. Never use them together with binoculars or a telescope.

As the Moon starts to move in front of the Sun you first see a small bite out of one edge. This slowly gets larger, and in a **partial eclipse** it doesn't move centrally across the Sun so it eventually moves slowly off again. This is the stage where you must view the Sun using an eclipse viewer or some other safe method.

But in a total solar eclipse the Moon passes centrally in front of the Sun, so it

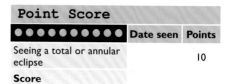

Point Score		
●●●●●●●●●●	Date seen	Points
Seeing a total or annular eclipse		10
Score		

Solar eclipses 2015–2024 (A = annular)	
Date	Where total
2015 Mar 20	N. Atlantic Ocean, Faroe Islands, Svalbard
2016 Mar 9	Indonesia, N. Pacific Ocean
2016 Sep 1	(A) Central Africa, Madagascar, Réunion, Indian Ocean
2017 Feb 26	(A) South America, S. Atlantic Ocean, Angola
2017 Aug 21	US states: WA, ID, WY, NE, KS, MO, KY, TN, NC, GA, SC
2019 Jul 2	S. Pacific, Chile, Argentina
2019 Dec 26	(A) Saudi Arabia, S. India, Sri Lanka, Indonesia
2020 Jun 21	(A) Central Africa, Saudia Arabia, Pakistan, India, China
2020 Dec 14	S. Pacific Ocean, Chile, Argentina, S. Atlantic Ocean
2021 Jun 10	(A) Canada, Greenland, Siberia
2021 Dec 4	Antarctica
2023 April 20	W. Australia, Indonesia
2023 Oct 14	(A) US states: OR, CA, NV, AZ, NM, TX; Central America, Colombia, Brazil
2024 Apr 8	Mexico; US states: TX, OK, AR, MO, IL, IN, OH, PA, NY, VT, NH, ME; Canada
2024 Oct 2	(A) S. Pacific, Chile, Argentina

blocks out the Sun completely for a few minutes. At this stage, it's quite OK to look directly at the Sun and enjoy the spectacle. Sunlight disappears from the landscape, and around the Moon's edge we see the Sun's pearly white atmosphere, called the corona.

If you're lucky you might see what look like pink flames at the edge as well. These are called *prominences* and are not the same as flames on Earth but are eruptions of hydrogen from the Sun. They do move, but too slowly to be seen during totality.

The brighter stars show up, as well as any planets, including those that happen to be too close to the Sun to be seen normally. Even on the hottest day the air gets cooler, and the light in the sky is like nothing you've ever seen before.

Because the distance of both the Sun and Moon varies, sometimes the Moon appears slightly smaller than the Sun, and doesn't completely cover it. In these cases a ring of bright light remains around the Moon's edge, in what is known as an *annular* eclipse. These are not as spectacular as a total eclipse, because the corona and prominences never become visible. You can't look at any stage of an annular eclipse directly – you must *always* use an eclipse viewer or other precautions.

After totality, it becomes another partial eclipse again. The whole event from beginning to end lasts two or three hours.

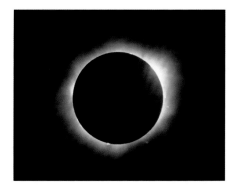

▲ At totality you can see the Sun's atmosphere and any bright prominences that happen to be taking place.

Partial eclipses are fun to watch as the Moon covers part of the Sun, so be well prepared for when the next one comes along!

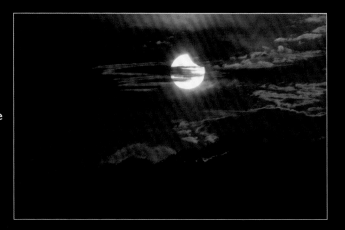

▶ A partial eclipse takes a bite out of the Sun. Always take care when observing one.

When to look

You are more likely to see a partial solar eclipse than a total solar eclipse, because they are seen from within the outer parts of the Moon's shadow when it crosses the Earth's surface, not from just the center. And there are some occasions when the center of the shadow misses Earth completely, so only a partial is visible from anywhere. Partial eclipses occur in the US and Canada during the eclipses of March 9, 2016 (HI and AK only); August 21, 2017; June 10, 2021 (northern and eastern mainland areas only); October 14, 2023; April 8, 2024; and October 2, 2024 (HI only) – see the list on page 97.

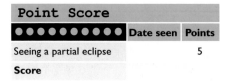

Point Score		
⬤⬤⬤⬤⬤⬤⬤⬤⬤⬤	Date seen	Points
Seeing a partial eclipse		5
Score		

What you'll see

Even with the Sun partly covered, it will still blind you, so you need the eclipse viewers mentioned on page 96 or some other way to view the Sun as in the **sunspots** article.

As the Moon starts to move in front of the Sun you first see a small bite out of one edge. This slowly gets larger and eventually covers a part of the Sun. Just how much it covers varies from eclipse to eclipse. Then the Moon moves slowly off again and after an hour or two the eclipse is all over.

If you can look at a partial eclipse in detail, you might be able to see that the Moon's edge is not completely smooth but can have bumps on it, caused by lunar mountains. And if the Moon's silhouette happens to cover a large sunspot, it appears darker than even the center of the spot. This is because the sunspot only appears dark compared with the rest of the Sun.

OBJECTS TO LOOK FOR IN WINTER

On the following pages you'll find objects that you can see mostly during winter.

The all-sky map on this page applies for February 1 at 8 pm, but you can recognize the same constellations a month or two on either side of that and earlier or later in the evening. The exact position of stars at the top and bottom of the map will vary depending on how far south or north you live.

The brilliant constellation of Orion is your guide – you really can't miss it when it's in the sky. A full description of Orion and the Orion Nebula (M42) is given as Object 36.

On the map are numbers that identify the constellations or objects you can look for. On those pages you'll get more help in finding each one.

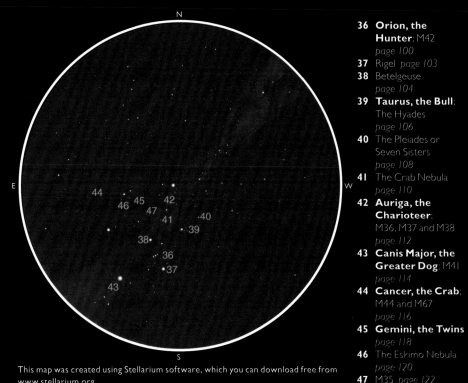

This map was created using Stellarium software, which you can download free from www.stellarium.org.

Orion, the Hunter
M42

Orion, the brightest constellation of all, rides high over winter nights and provides a signpost to many other constellations.

Where to look

Orion is visible throughout the winter, starting in the late evenings in October, when it is rising in the east, and being most obvious in February when it's centrally placed in the middle of the southern sky. By May it is setting over in the west not long after sunset. It is equally visible from both northern and southern hemispheres, though people in the southern hemisphere see it in their summer. You can always recognize Orion by the three bright stars in a straight line that make up Orion's Belt.

Legends

Orion is the Hunter, and it is a personal name rather than being Latin for anything. There are many myths and legends surrounding him, and maybe every folk tale that required a great hunter used this most splendid of constellations to represent him. Early astronomical records in both Mesopotamia (now Iraq) and China show these stars as a hunter. In the skies he is fearlessly facing the charging bull Taurus, there is another meal in the shape of a hare, Lepus, just below him, and he is followed by two faithful dogs, which we call Canis Major and Canis Minor.

In one story, he took a fancy to Pleione, the mother of the **Pleiades** (the Seven

Point Score		
⚫⚫⚫⚫⚫⚫⚫⚫⚫⚫	Date seen	Points
Seeing Orion		I
Identifying the bright stars		I
Seeing M42		2
Seeing Trapezium as four stars		3
Score		

See also: **Betelgeuse; Rigel.**

Sisters star cluster in Taurus). Zeus, king of all the gods, took a dim view of this and put them all in the sky, where Orion continues to pursue Pleione and her daughters to this day.

What you'll see

The figure of Orion is one of the most recognizable in the whole sky. Three bright stars in a slanting line are flanked by four other stars. The stars in a line are the belt of the Hunter, who faces Taurus, the Bull, and the four stars mark parts of his body. The lower right star, **Rigel**, is his upraised foot and the top left star, **Betelgeuse**, is his shoulder. Fainter stars depict his head, a shield that he has raised against the bull, and a raised club.

To the left of Orion, also facing the bull, are his faithful dogs, one large and one small – **Canis Major** and Canis Minor. Below him is Lepus, the Hare, though maybe the charging bull has grabbed the dogs' attention and he will escape unharmed.

Most of the main stars of Orion are also young, blue stars, and are associated with each other, having formed from a giant cloud in that region of space. Star formation is still going on

▲ The main stars of Orion, with the Orion Nebula showing pink below the three Belt stars, though you won't see this color visually.

Fact File

Name   Orion, the Hunter (name of legendary character)
Area   594 square degrees

Objects		Magnitude	Distance	Type	Visibility
Betelgeuse	Alpha Orionis	0.4	860 light years	–	–
Rigel	Beta Orionis	0.1	497 light years	–	–
Bellatrix	Gamma Orionis	1.6	252 light years	–	–
Mintaka	Delta Orionis	2.4	690 light years	–	–
Alnilam	Epsilon Orionis	1.7	1,970 light years	–	–
Saiph	Kappa Orionis	2.0	245 light years	–	–
Orion Nebula, M42		–	1,600 light years	Gaseous nebula	A

in the area, around the Orion Nebula.

Also known by its catalog number M42, the Orion Nebula is one of the brightest nebulae in the sky. Nebula is a Latin word meaning "cloud," and its Latin plural is "nebulae," usually pronounced "neb-you-lee," though some books think people can't cope with this and put "nebulas." (They use "supernovas" as well.) You can see the Orion Nebula for yourself. It is bright enough to be seen with the naked eye in dark, country skies, but most people will get a better view using binoculars or a telescope. To find it, look in the middle of

Brain Box

The top right star of Orion is called Bellatrix, from which J. K. Rowling took the name of her character Bellatrix Lestrange for the Harry Potter books. The star Bellatrix is an imposter – it is closer and not one of the Orion group of stars.

the area between Orion's Belt, Rigel and Saiph, which is the lower left-hand star of the four surrounding the Belt. There is a line of three fainter stars, and the Orion Nebula surrounds the middle one.

In city skies you may only notice that it is slightly hazy, but in good conditions it starts to look a bit like the photos, though not in color. You can see the wing-like structures that extend away from the middle, and a darker sort of bay known as the Fish's Mouth. Beyond this is a fainter area known as M43.

With a telescope, look closely at the star in the center of M42 and you should see that it is in fact a tight group of four stars, known as the Trapezium from the shape of the group.

▼ A digital camera shot of the Orion Nebula through a 200 mm telescope shows the spectacular swirls of hydrogen gas with the Trapezium in the center.

Rigel

37

Brilliant, blue and massive, Rigel is not only the brightest star in Orion, but it is one of the most luminous stars in our part of the Milky Way.

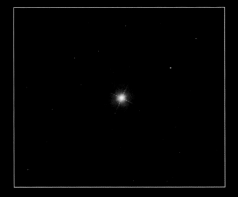

▶ Rigel is an obvious blue color. The spikes on this picture are due to the telescope, not the star.

Where to look

Rigel is the lower right-hand star of the four that surround Orion's Belt. It is usually the brightest of the four, and, like Orion, is visible during evenings from October to May. In the mythological figure, Orion traditionally has his left leg raised as he advances toward the charging bull, Taurus, and Rigel marks his left foot. (Use the finder chart on page 100.)

What you'll see

Like all stars, no matter how large, Rigel appears as just a point of light as seen through binoculars or a telescope. But you will see its blue color particularly strongly using binoculars. It might not be strongly blue, but as stars go this is quite obvious.

Point Score

●●●●●●●●●●	Date seen	Points
Seeing Rigel		1
Seeing Rigel B		7
Score		

Rigel has a nearby companion star, called Rigel B, but you are unlikely to see this unless you have both a telescope and very steady conditions. It is quite close to Rigel, so it is hard to pick out against the glare of the main star. At magnitude 6.8, Rigel B is not particularly faint, and it is separated from Rigel by 10 arc seconds, which even the smallest telescope has enough power to show. But Earth's unsteady atmosphere usually makes Rigel twinkle, so that it often looks like a moving blob rather than a point of light. You will need a magnification of 100 or more.

Soundbite

The name is pronounced to rhyme with "Nigel."

Fact File

Name	Rigel (Beta Orionis)
Magnitude	0.12
Distance	497 light years
Mass	18 times that of the Sun
True brightness	120,000 times that of the Sun
Diameter	About 200 million km

This is one of the most famous stars in the sky, but there are suggestions that it could actually explode at any time!

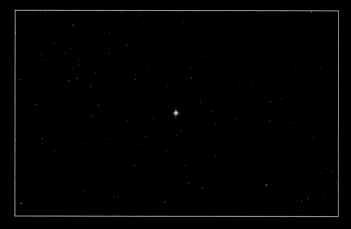

◀ Betelgeuse's red color shows up particularly well through a telescope or binoculars.

Where to look

Betelgeuse is the top left-hand star of the four that make up the body of the constellation of Orion, the Hunter. It is one of the brightest stars in this part of the sky, and has a noticeable reddish tinge. (Use the finder chart on page 100.)

Brain Box

How do you pronounce its name? Most astronomers say something like "Bet-el-jooz" but many people just go for "Beetle juice." The 1980s film featuring a ghost with this name had nothing to do with the star.

What you'll see

You might think that one star looks very much like another, and they are all white, but Betelgeuse is definitely a bit different. You can see that it is reddish straight away, particularly if you compare it with Rigel at the bottom right of Orion, which is bluish. Binoculars will help to bring out the color, and you won't be surprised to hear that Betelgeuse is a red supergiant star.

This is a massive star, around 20 times the mass of the Sun, and shining about 140,000 times as bright. Not only is it massive, it is huge. Red giant and supergiant stars swell up as they approach the ends of their lives, and in the case of Betelgeuse

Fact File

Name	Betelgeuse (Alpha Orionis)
Magnitude	0.42
Distance	860 light years
Mass	20 times that of the Sun
True brightness	140,000 times that of the Sun
Diameter	800 million km, give or take lots of millions

its diameter may be as large as the orbit of Mars around the Sun.

But don't expect to see a big star through your telescope. At nearly 500 light years away, even Betelgeuse appears as only a point of light unless you have access to some very special kit, such as the Hubble Space Telescope, which has taken a rather blurry picture of the star. It shows that Betelgeuse is not completely circular, and it probably even changes its shape over a period of time. It seems to have a rather loose grip on its outer regions. The amazing thing is that despite being enormous and very bright, the outer regions of the star are by Earth standards a vacuum, with the gas being far less dense than our upper atmosphere. So its actual diameter is very hard to pin down.

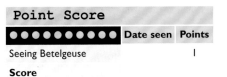

Point Score		
⬤⬤⬤⬤⬤⬤⬤⬤⬤⬤	**Date seen**	**Points**
Seeing Betelgeuse		I
Score		

Nor is Betelgeuse always the same brightness. Usually it is slightly fainter than Rigel, but occasionally it becomes the brightest star in Orion. So compare the two from time to time to see what's happening.

Eventually, sometime within the next million years or so, Betelgeuse will explode as a supernova (see page 111). This has led to newspaper headlines saying that "Betelgeuse could explode tomorrow." Yes, it could, but then there's an equal chance that it will explode in a million years' time as far as we know!

◄ The Hubble Space Telescope shows that Betelgeuse doesn't have an even brightness all over, though even the HST can't see real detail on the surface.

39

Taurus, the Bull
The Hyades

Taurus contains the two most easily seen star clusters in the whole sky – the Hyades and the Pleiades.

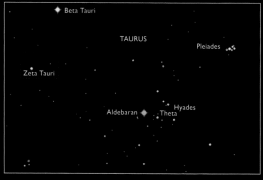

Where to look

Taurus is a particular feature of winter nights, but you can see it from October to April evenings. If you can find **Orion**, you can find Taurus. Follow the three stars of Orion's Belt upward to the northwest, and you come to a bright star called Aldebaran. This marks the eye of the bull, and it's surrounded by the first of our clusters, the Hyades. Follow the line a little farther and you get to the Pleiades.

Going back to the Hyades, you'll notice that they have a V-shape, which is thought of as the face of the bull. Follow the arms of the "V" upward and you come to two other stars marking the tips of its horns. This is no modern bull with short horns, but one with really long horns. However, in the sky the hind quarters of the poor creature are cut off to make way for other constellations.

Legends

The bull in the heavenly Greek myths is nothing more than a disguise for the chief god, Zeus. You might think that all the mythical misses would fall for Zeus straight away, but they were much more streetwise than that even in those legendary times and he had to resort to various deceptions to seduce them.

Point Score		
⬤⬤⬤⬤⬤⬤⬤⬤⬤⬤	Date seen	Points
Finding main stars of Taurus		I
Finding the Hyades		I
Finding Theta Tauri		I
Score		

See also: **Pleiades; Crab Nebula.**

One such was a lovely lass called Europa, who came across a beautiful but amiable bull on the seashore while she was playing with her friends. She was delighted when she found she could actually ride the bull into the surf, but she soon discovered the error of her ways when he carried her off to a distant shore. We will draw a modest veil over what happened next, but this maritime escapade is said to be the explanation for only the bull's front half being shown in the sky – the rest of him is hidden beneath the waves.

Brain Box

Aldebaran itself is not actually part of the Hyades cluster, as it is only 66 light years away. It is a red giant star, and varies slightly in brightness. See the box on page 145 for more about red giants.

What you'll see

You can see the individual stars of the Hyades with the unaided eye, but binoculars will show them better if your skies are poor. Look for a pair of stars together called Theta Tauri, which is one of the brightest stars in the area. Each of these stars is about 65 times brighter than the Sun, but at the distance of the Hyades, about 152 light years, they seem quite dim. The dimmer is called Theta 1 and the brighter (the more southerly) is Theta 2.

▼ A close-up of the Hyades, with red giant Aldebaran.

Fact File

Name	Taurus, the Bull			
Area	797 square degrees			
Objects			Magnitude	Distance
Aldebaran	Alpha Tauri		1.0 (variable)	66 light years
Elnath	Beta Tauri		1.7	134 light years
	Zeta Tauri		3.0	444 light years

Soundbite

The word Hyades is pronounced "Hy-adeez."

◄ Taurus represents the front end of a charging bull.

40

The Pleiades or Seven Sisters

This is the best and most famous star cluster of all. But why are there nine named stars in the Seven Sisters?

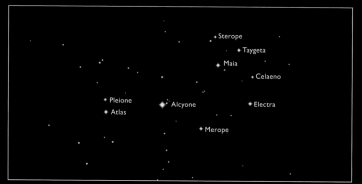

◀ The nine named Pleiades. Sterope is an alternative spelling of Asterope.

Where to look

First find the constellation of **Taurus**, which is visible from autumn to spring, then look a little to the northwest. In a good sky you'll spot this little cluster of stars easily, but even in poor conditions they show up as a fuzzy patch. Although none of the stars is particularly bright, together they attract the eye and you can't miss them. The cluster also has the catalog number M45. (Use the finder chart on page 106.)

Point Score		
⚫⚫⚫⚫⚫⚫⚫⚫⚫⚫	Date seen	Points
Finding the Pleiades		1
Counting at least six stars with the naked eye		2
Score		

Legends

The Pleiades were seven sisters who were the daughters of a god named Atlas and a water goddess, Pleione. Atlas, incidentally, had the job of holding the heavens aloft, and a picture of him holding up a globe was often used on collections of maps, which therefore became known by the same name.

An individual sister is known as a Pleiad ("Ply-ad"). Their names are Alcyone, Asterope, Celaeno, Electra, Maia, Merope and Taygeta. In the sky, they are joined by their parents, which is why nine of the stars have names.

What you'll see

How many Pleiades you can see depends on how good are your skies, and your eyes. Many people can see only six rather than seven or nine, because Asterope and Celaeno are a bit fainter than the others, and Pleione is fairly close to her husband Atlas so you might not spot her. But people with good eyesight can see more than nine. If you use binoculars you can see dozens, and there are about 100 stars there altogether.

Star clusters are groups of stars that formed together. When stars form out of gas and dust, many are usually born at around the same time. Over millions of years they drift apart and become separated, but when they are comparatively young they can still be seen close together. The Pleiades cluster formed about 100 million years ago, and still contains

Soundbite
The word Pleiades is pronounced "Ply-adeez."

some of the bright, blue stars that are a sign of a young cluster.

Photographs show a blue mist around some of the stars. They happen to be passing through a dusty area of the galaxy, and the dust reflects the blue light of the brightest stars.

Fact File

Name	Pleiades (Seven Sisters), M45
Type	Open cluster
Visibility	A

Individual stars	Magnitude	Distance
Alcyone	2.9	402 light years
Atlas	3.6	404 light years
Electra	3.7	404 light years
Maia	3.9	382 light years
Merope	4.2	379 light years
Taygeta	4.3	408 light years
Pleione	5.0	381 light years
Celaeno	5.6	376 light years
Asterope	5.8	371 light years

▼ A long-exposure photo reveals veils of dust surrounding these modest Seven Sisters (and their parents).

It has a great name and a great story behind it – but the Crab is quite a tricky beast to catch because of its faintness.

Crab Nebula

Zeta Tauri

▶ This photo of the Crab Nebula (M1) through an 80 mm refractor gives an idea of how it looks through a telescope. The star at lower left is Zeta Tauri.

Where to look

Back in the year 1054 there was no doubt where to look – there was a brilliant new star in Taurus, visible even in daylight. But over the months it faded and now all we can see is the remains of the star that blew itself up. To find it now, start with **Taurus** and find the lower horn of the bull, which ends in the star known as Zeta Tauri. Just to the northwest (upper right) of this star lies the Crab Nebula (M1). It will probably not be visible in the finder telescope, if you have one, so all you can do is aim the telescope carefully at the right place and hope that you can spot it in the main telescope.

Observing tip

If you think you're looking in the right place but can't see it, try moving the telescope around a little. It's often easier to spot a faint moving object than a stationary one. Also try looking anywhere in the field of view other than where you expect it to be. The edge of your vision is more sensitive to light than the center, and deliberately looking a little away from the spot is called *averted vision*.

It's large enough to be seen with binoculars as well, but only if you have a very clear sky, or large binoculars.

What you'll see

The Crab Nebula, once so brilliant, now shows itself only as a very pale, hazy, oval blur. It's about a fifth of the width of the Moon, so remember how big the full Moon is through your telescope or binoculars

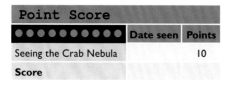

Point Score		Date seen	Points
●●●●●●●●●●			
Seeing the Crab Nebula			10
Score			

Supernovae

A supernova is basically a star that explodes at the end of its life, but not every star becomes a supernova. It is only the most massive stars, typically more than about 10 times as massive as the Sun, which do this. What happens is that the nuclear reactions inside the star stop working because the fuel runs out, so the whole star collapses down on itself and explodes, scattering most of its material far and wide. At the time of the explosion, the star can be as bright as billions of stars put together for a few weeks.

Smaller stars can also collapse, and sometimes they explode in a supernova as well, but only if there is another star very close by which dumps extra material on to them. The plural of supernova is usually supernovae, pronounced "super-no-vee," but sometimes they are called "supernovas."

▼ The remains of the exploded star look spectacular in a photo with a 200 mm reflector, but don't expect to see it looking like this through even a large telescope.

and look for something about a fifth of its diameter.

Photographs show a mass of tangled strings or filaments of gas with a pair of stars at their center. One of these stars has nothing to do with the Crab Nebula, and is just in the line of sight, but the other is the mortal remains of the star that was seen to explode in 1054 in what's now called a supernova – one of the greatest explosions that can occur. For more on these stars, check out the adjacent box. But the stars at its center are too faint to be seen other than in really large telescopes. The Crab Nebula is the first in Messier's catalog of deep-sky objects (see page 131), so it is also referred to as M1.

Fact File

Name	Crab Nebula, M1		
Type		Visibility	Distance
Supernova remnant		D	6,000 light years

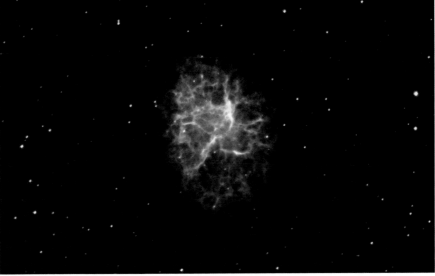

42

Auriga, the Charioteer
M36, M37, M38

It contains one of the brightest stars in the sky, yet Auriga is not as well known as any of its neighbors.

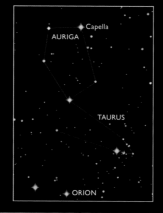

Where to look

Immediately above the winter constellations of Taurus and Orion lies Auriga. To find it, start from Orion and draw a line upward through the middle of the constellation. You will come to a large pentagon of stars, with its brightest star, Capella, at the top. Auriga is virtually overhead on January and February evenings. Actually, the bottom star of the pentagon is in Taurus, but it helps to make the Auriga pattern and was once called Gamma Aurigae. Today, however, the shared star is known only as Beta Tauri and there is no Gamma Aurigae.

Legends

Auriga is a name rather than a translated word, but there is no Greek hero with this name. Instead, some legends say that he

Point Score			
○○○○○○○○○○		Date seen	Points
Finding Auriga			1
Finding Haedi			2
Finding M36			3
Finding M37			2
Finding M38			2
Score			

is really Erichthonius – let's call him Eric for short – a king of Athens who invented the four-horse chariot. This clever idea was rewarded by Zeus, the king of the gods, who placed Eric among the stars. But the odd thing is that there is no chariot ever shown in the drawings of Auriga, the Charioteer, and instead the poor chap is carrying a she-goat and her two kids. Two separate legends have become mixed up over the years, but you can still see the kids in the stars.

What you'll see

Capella, Auriga's brightest star, is a yellow star, so if you want to be poetic you can call it golden. As the sixth brightest star in the night sky it really does glisten. Just to the right of Capella is a triangle of three stars of similar brightness, around third magnitude. The lower two of these are known as the Haedi, which is Latin for "Kids."

▲ Each of these clusters, M36, M37 and M38, is slightly different from the others.

Fact File

Name	Auriga, the Charioteer				
Area	657 square degrees				
Objects		Magnitude	Distance	Type	Visibility
Capella	Alpha Aurigae	0.1	43 light years	–	–
Menkalinan	Beta Aurigae	1.9	40 light years	–	–
M36		–	4,100 light years	Open cluster	C
M37		–	4,400 light years	Open cluster	B
M38		–	4,200 light years	Open cluster	B

The other features of interest in Auriga are three star clusters more or less in a line, named M36, M37 and M38, though they don't occur in that order in the sky. They are a good test of your sky and your eyesight. In a really dark and clear sky they are easily visible with the naked eye as three little hazy patches, more or less evenly spaced, at right angles to the line between Beta Tauri and Beta Aurigae. With binoculars you can see them more clearly, but they still look like misty patches rather than star clusters. From suburbs you might just see them with binoculars, but this depends on just how dark the sky is. Not until you look with a telescope can you see that they are composed of quite faint stars, and in city skies you might have to wait for the very clearest night to see these clusters.

Each cluster is slightly different. M37, the one at the left, has the most stars, but they are fainter than those in the center one, M36, which is the smallest cluster. In poor skies this might be the only one you can see. The one on the right, M38, is the largest. Actually, no two star clusters are the same, so you might like to compare these with others in the same region of sky, such as **M35** in Gemini and the **Double Cluster** in Perseus.

43

Canis Major, the Greater Dog M41

Orion's faithful dog follows him through the sky, with the brightest star, Sirius, the Dog Star, in his mouth.

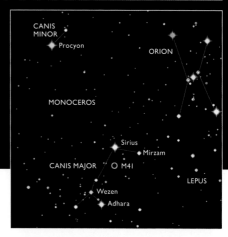

Where to look

This is one of the constellations that **Orion** conveniently points out to us. Follow the line of the three stars of Orion's Belt to the southwest and you come to Sirius, unmistakable as it is the brightest star in the night sky. The other stars of Canis Major extend from Sirius downward. It is fairly far south, so those in northern parts will have to choose their time. It is visible from late evenings in November until the early evening in May.

Legends

Which particular dog is represented by Canis Major depends on which Ancient Greek storyteller you listened to. He is always the fastest and most able hunter, and originally Sirius itself was the dog rather than the whole constellation. Maybe the link between Sirius and a dog comes because Orion was the hunter, and Sirius always follows Orion through the sky.

Point Score			Date seen	Points
Seeing Canis Major				1
Seeing M41				2
Score				

In some stories he is called Laelaps. He was a dog that could always catch his prey, yet on one occasion he was set to catch a fox that was completely uncatchable. The chase still continues, but there is no sign of the fox in the winter sky, though there is one in summer, Vulpecula. However, it sets just as Canis Major rises, and appears on the other side of the sky when Canis Major disappears in spring, faithfully following Orion. The only trouble with this story is that the constellation of Vulpecula was not invented in Ancient Greek times, but let's not allow the facts to stand in the way of a good tale.

What you'll see

Sirius itself dominates the constellation. When it is low in the sky, it often twinkles madly. Twinkling of stars is caused by Earth's atmosphere, and on some nights it is much more obvious than on others. Sometimes, twinkling Sirius even seems to be changing color from its usual white, and to be shooting out colored rays in all directions. People have been known to report it as a UFO. Twinkling is a dramatic sight in a telescope, but you are not looking at anything astronomical.

It is the second closest bright star to Earth, just about twice as far away as Alpha Centauri, which is a southern star.

Canis Major also contains a bright star cluster, M41. This is almost directly south of Sirius, and is visible to the naked eye on good nights, though if you have a poor sky and the constellation is low down you might have trouble even with binoculars. Binoculars or a telescope will show upward of 50 stars in an area the size of the Moon.

Also look for

There is a Lesser Dog, Canis Minor, as well, marked by the bright star Procyon, some way north of Canis Major. There is nothing much of interest in Canis Minor, though it has a second-magnitude star, Gomeisa, fairly close by. Sirius, Procyon and Betelgeuse in Orion are known together as the Winter Triangle.

▼ The cluster M41 is directly south of Sirius, the brightest star in the sky.

Fact File

Name Canis Major, the Greater Dog
Area 380 square degrees

Objects		Magnitude	Distance	Type	Visibility
Sirius	Alpha Canis Majoris	−1.5	8.58 light years	–	–
Mirzam	Beta Canis Majoris	2.0	491 light years	–	–
Wezen	Delta Canis Majoris	1.8	1,600 light years	–	–
Adhara	Epsilon Canis Majoris	1.5	400 light years	–	–
M41		–	2,300 light years	Open cluster	B

44

Cancer, the Crab
M44 and M67

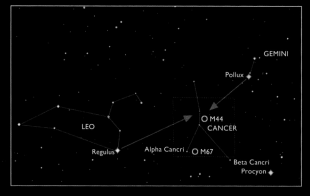

Though everyone has heard of this constellation in the Zodiac, it is quite hard to find – but it's well worth the effort.

Where to look

With some constellations you need little help to find them, but Cancer is one that you could have trouble with if there's a lot of light pollution. Even its brightest stars are only fourth magnitude, which are hard to see in bright city skies. It is a winter and spring constellation, and lies between **Leo** and **Gemini**. Perhaps the best way to spot it is to look midway between the bright stars Pollux in Gemini and Regulus in Leo. That takes you to roughly the heart of the crab, where there are two fourth-magnitude stars one above the other. From here you can pick out the other stars. Alpha and Beta Cancri lie along a line between Regulus and Procyon, the brightest star in Canis Minor.

Legends

Surprisingly for a constellation in the Zodiac, the Crab plays only a minor role in Greek myths. Its star role was to pester Hercules, also known as Heracles, and for its trouble was stamped on. So quite why this part of the sky has been associated with such a lowly creature, placed on equal footing with others such as lions and bulls, is a bit of a mystery.

What you'll see

For many people Cancer is one and the same with the star cluster M44, which is one of the most easily visible clusters in the sky. Though not nearly as bright as the **Pleiades**, it is one of the few star clusters that can be seen from city centers with binoculars, and is very easy to spot

Point Score		
●●●●●●●●●●	Date seen	Points
Seeing Cancer		2
Seeing M44		2
Seeing M67		3
Score		

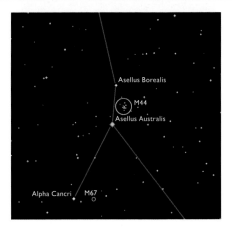

M44 has two other names – the Beehive, based on its general appearance, and Praesepe (pronounced "Pry-seepee"), which is Latin for "manger." The Ancient Greeks referred to the two stars as the two asses, and this lives on in their names – Asellus Borealis and Asellus Australis.

Not far away in Cancer is another cluster suitable for binoculars and telescopes, just plain M67 with no other names. Find it just to the right of the star Alpha Cancri. It will be in the same field of view using binoculars, but with a telescope you'll have to move a couple of fields of view due west. It is much smaller and fainter than M44, but has plenty of stars when viewed through a telescope.

with the naked eye from darker areas. But though they are quite widely spaced out, none of its stars are bright enough to be seen with the naked eye, so all you will see is quite a large misty patch.

However, even the smallest of binoculars should reveal many stars in this area. There are two or three dozen bright enough to be seen easily. This is a case where binoculars are just as good as a telescope, because the stars cover about a degree of sky. Many telescopes will magnify too much, giving a less dramatic view. But from a suburban location, this is a good object to show off your telescope. Dozens of stars appear in a patch of sky that looks like any other.

▼ The main stars of Cancer, with the Beehive Cluster. You can also just see M67 near the bottom of the picture, to the right of Alpha Cancri.

Fact File

Name	Cancer, the Crab				
Area	560 square degrees				

Objects		Magnitude	Distance	Type	Visibility
Acubens	Alpha Cancri	4.3	188 light years	–	–
	Beta Cancri	3.52	303 light years	–	–
Asellus Borealis	Gamma Cancri	4.7	180 light years	–	–
Asellus Australis	Delta Cancri	3.9	130 light years	–	–
M44 (Praesepe or Beehive)		–	600 light years	Open cluster	A
M67		–	3,000 light years	Open cluster	C

45

Gemini, the Twins

The stars Castor and Pollux are often called the Heavenly Twins, but the figures of the twins are shown in the sky as well.

Where to look

Gemini is a winter and spring constellation, though it begins to become visible in late evenings over in the east around the start of October and remains visible until May, when it sets in the west just after the Sun. Look for a pair of stars of roughly equal brightness, and about a fist's width apart, above and to the left of Orion.

Legends

These twins both had the same mother, Leda, but different fathers. As a result, Castor was mortal but Pollux was immortal. The twins went through many adventures, but eventually the inevitable took place and Castor was killed. His immortal brother, who had fought so many battles by his side, was so grief-stricken that his father, Zeus, decided to give both

Brain Box
It's easy to forget which Twin is which. But if you know the rest of the sky, there's a quick way to remember. Castor is closest in the sky to Capella, while Pollux is closest to Procyon. All very well as long as you can remember which are Capella and Procyon!

of them a place among the stars, so Castor now lives in the heavens forever.

What you'll see

The two main stars, Castor and Pollux, mark the heads of the Twins, but their bodies are also marked by lines of stars. Castor's feet are marked by a fainter and closer pair of stars, Mu and Eta Geminorum, while those of Pollux are shown by Alhena, which is almost as bright as Castor. Pick out the two lines of stars to get to know the shape of the constellation.

Point Score		
●●●●●●●●●	Date seen	Points
Finding Castor and Pollux		1
Finding the lines of stars marking each Twin		2
Score		

Castor is an interesting star. Quite a small telescope shows that it is a double star, with one of the stars being less than half the brightness of the other. The two main stars orbit each other, but don't expect to see any changes from night to night. The orbital time – known as the period – is 467 years. So even in a human lifetime you would only see a small change in their appearance. Close by is another much fainter star, known as YY Geminorum, which takes thousands of years to orbit the main pair of stars. And, curiously, all of these stars are double or binary stars, though the companions are so close in each case that they can't be distinguished from them, with periods of a matter of days. So Castor is actually six stars, not one.

Pollux, by comparison, is a single star, though it is cooler and redder than the stars that make up Castor. It is also brighter, so really the Heavenly Twins are not twins at all.

▲ Castor as it appears through a small telescope. As well as the main double star, look for ninth-magnitude YY Gem nearby.

▼ The Twins rise in the east on late November evenings.

Fact File

Name	Gemini, the Twins		
Area	514 square degrees		
Objects		Magnitude	Distance
Castor	Alpha Geminorum	1.6	51 light years
Pollux	Beta Geminorum	1.2	34 light years
Alhena	Gamma Geminorum	1.9	30 light years

46

The Eskimo Nebula

Political correctness has yet to catch up with the Eskimo Nebula, a bright planetary nebula in Gemini.

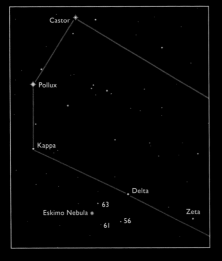

▲ A photo taken with a 200 mm telescope shows that the Eskimo has a quite starlike center and a blue gas shell around it.

Where to look

The Eskimo Nebula is fun to find using the technique of star-hopping, which means starting from some point you can find easily and recognizing nearby star patterns, hopping from one to the other. In this case, you will use your telescope's finder to begin with.

Start with Pollux in **Gemini**. From there, find Kappa Geminorum, which is about the same distance as Castor, but due south of Pollux, then find Delta Gem, which is the same brightness as Kappa but the next star along down Pollux's body, going toward Alhena at his foot.

There are two fainter stars forming a triangle with Delta, 63 and 56 Gem, of which you want the one to the east of Delta, which is 63 Gem. From here on it

gets a little tricky, as the Eskimo is quite faint and small. South of 63 Gem is 61 Gem, and the Eskimo Nebula makes a triangle with 63 and 61 Gem. Use a low-magnification eyepiece first, as always, but don't expect to see the Eskimo straight away as it is rather small. Look for a pair of fairly faint stars close together. One of them is an ordinary eighth-magnitude star, but the one next to it is the Eskimo Nebula, and is fainter though you might notice that it seems slightly fuzzy. Once you've found it,

Point Score		
●●●●●●●●●	**Date seen**	**Points**
Finding the Eskimo Nebula		5
Making a drawing		3
Score		

you can increase the magnification, ideally to around 100 or more.

If this method doesn't work for you, there is a slower way to find the right part of the sky for the Eskimo Nebula. Locate the star Zeta Geminorum, which through a telescope has a fainter star close to it. Get it to the middle of the field of view of your lowest-power eyepiece, then leave the telescope (with motor drive, if you have one, switched off) for exactly 25 minutes without touching it. When you come back, the pair of stars, one of which is the Eskimo Nebula, will be just slightly north of center in the field of view, courtesy of the Earth turning by just the right amount!

What you'll see

To start with, probably very little! The Eskimo Nebula is about the same size as Jupiter as seen in the telescope, but with the lowest magnification this is not very large. The central part of the Eskimo is the brightest, so if your sky is not very dark you may only see this bit and not the fainter outer regions. In most cases, increasing the power to around 100 or even more is a good idea.

You can see the central star quite easily, but seeing more of the nebula surrounding it

▶ As seen with the Hubble Space Telescope, the Eskimo comes into its own with a fur parka surrounding a sort of face.

calls for a bit of technique. Use averted vision and you may well see more. There's a brighter region of nebula immediately around the star, and a larger but fainter area surrounding that.

Can you see any color? Planetary nebulae can often be blue or green, and whether you see the color depends partly on how big your telescope is (the bigger the better) and on your eyesight.

Viewed with a large telescope, the outer parts of the Eskimo Nebula show streaks and there is mottling in the central area. People saw the appearance as a bit like the face of an Eskimo surrounded by a fur parka – but that was in the days before the Canadian Inuit people objected to being called Eskimos. An alternative name is the Clown Face Nebula.

Fact File

Name Eskimo Nebula, NGC 2392		
Type	**Visibility**	**Distance**
Planetary nebula	C	3,000 light years

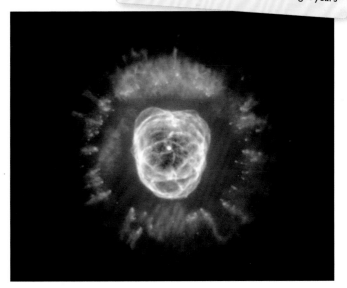

A rich star cluster in Gemini that is easy to spot with binoculars or even the naked eye under the right conditions.

▶ M35 is one of the easiest star clusters to find, though you need good skies to see it looking like this. Use the map on page 118 to help you find it.

Where to look

Having found Gemini, go to the foot of Twin Castor, where there are two stars, Mu and Eta Geminorum. Look an equal distance to the separation of these stars to the right of Eta, but at an angle to the line between them, and you should spot the cluster M35. Often in the sky you have to make angles and triangles with known stars to find objects, so finding M35 is good practice for more challenging objects.

What you'll see

Because M35 is quite a large cluster, even binoculars show individual stars in the group rather than just a general haze, as happens with smaller or fainter clusters.

Fact File

Name	M35		
Type		Visibility	Distance
Open cluster		B	2,800 light years

None of the stars are particularly bright, but as seen with binoculars just a few of them will cut through the light pollution. Even from a city sky you should see just a few pinpricks of light against the orange glow, and from a darker site the cluster gets better and better.

There are about 50 stars here visible with binoculars in an area about the same size as the full Moon in the sky. One or two of these are actually not members of the cluster but just happen to be in the line of sight. The cluster itself is quite distant, in a neighboring spiral arm of our galaxy. All the stars you can see with binoculars in M35 are very much brighter than the Sun. You would need at least a 300 mm telescope to see a star like the Sun in M35.

Point Score

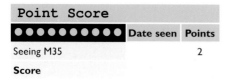

	Date seen	Points
Seeing M35		2
Score		

OBJECTS TO LOOK FOR IN SPRING

The following pages show objects that appear in the skies of spring.

In spring it gets dark later, so the all-sky map on this page applies for May 1 at 10 pm but you can recognize the same constellations a month or two on either side of that and later in the evening. The exact position of stars at the top and bottom of the map will vary depending on how far south or north you live.

The numbers on the map show you where to find the objects that are shown in detail on the following pages. Get your bearings by looking for the crouching lion shape of Leo, in mid sky, and also by following the curve of the handle of the Big Dipper down toward the horizon to find the bright stars Arcturus and then Spica.

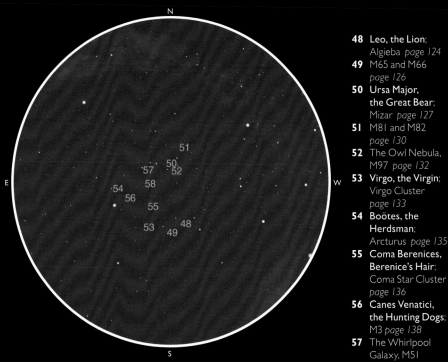

This map was created using Stellarium software, which you can download free from www.stellarium.org.

Leo, the Lion
Algieba

A key constellation of the spring sky, the stars of Leo make a pattern that actually resembles what it is meant to be.

Where to look

From December to July, Leo is in the evening sky. It rises late on December evenings, then dominates the mid sky during the spring. It finally sinks into the western twilight in July. Its brightest star, Regulus, stands out to the east of **Gemini** and marks the heart of the lion, but the other stars that make up the pattern are not particularly bright. Even so, the outline of a crouching lion with its fine mane is easily spotted.

The curve of stars that makes up the mane is also known as the Sickle, from its shape. Several stars make up the body of the lion, and its rump is marked by Denebola, from an Arabic word meaning "lion's tail."

Legends

Greek myths said that Leo was a fearsome lion that could not be killed until Heracles, whom we also know by his Latin name of Hercules, came on the scene. Such was his strength that he was able to strangle the beast. However, the association of these stars with a lion probably dates back beyond Ancient Greek times.

What you'll see

The shape of the lion is quite easy to pick out. But finding other objects to view takes a little work. There are no star clusters here, as Leo is well away from the Milky Way region. Instead, we have a bright double star and galaxies **M65 and M66**.

Point Score		
●●●●●●●●●●	**Date seen**	**Points**
Finding Leo		1
Identifying Regulus, Denebola and Algieba		1 each
Seeing Algieba as a double star		3
Score		

▶ Algieba is a double star with two orange stars of almost equal brightness.

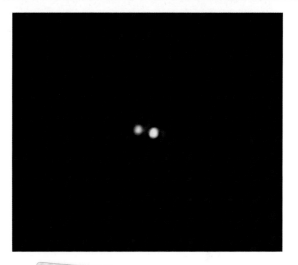

The double star is Algieba, also known as Gamma Leonis. Through binoculars it is just a single star, but a telescope with a high magnification shows that it is a double star. Unlike many doubles which are plain old white, the two stars of Algieba are orange in color, making it an attractive sight. But you will need both a high magnification – 150 or 200 – and a steady atmosphere, as the two stars are quite close together at 4 arc seconds. Under choppy seeing, they can merge together and you'll see only a writhing blur.

▼ The stars of Leo fit the pattern of a crouching lion very well.

Fact File

Name	Leo, the Lion		
Area	947 square degrees		
Objects		Magnitude	Distance
Regulus	Alpha Leonis	1.4	79 light years
Denebola	Beta Leonis	2.1	91 light years
Algieba	Gamma Leonis	2.0	130 light years

▲ M66 (lower left) and M65 are both spiral galaxies at an angle to us. NGC 3628, which is harder to see, is almost edge-on so its bright nucleus is hidden by a dust lane.

Two distant galaxies in Leo that provide a nice challenge for binoculars and small telescopes.

Where to look

Begin with Denebola in Leo, then move one brightish star toward Regulus, which is Theta Leonis. From there, find Iota Leonis, which makes a large triangle with Denebola and Theta. The two galaxies M65 and M66 are exactly midway between Theta and Iota Leonis. With typical binoculars with a 5° field of view, put Theta at one edge of the field of view and Iota at the opposite edge, and the two galaxies should be in the middle, to the left of an arc of stars.

They are quite tiny, so don't expect them to jump out at you, particularly if you are using binoculars. Wait until Leo is as high

as possible – which means when it is due south – and don't even bother looking if conditions are slightly misty. You need a sky which is as dark and clear as you can get.

What you'll see

The galaxies show up as two small, slightly oval misty patches. The darker your sky, the larger they will appear. What you are seeing are the centers of the galaxies rather than the spiral arms which show up on long-exposure photographs. The easternmost galaxy, M66, is the brighter of the two, but M65 is more noticeably spindle-shaped.

Just to the north is a fainter and larger galaxy, NGC 3628. This is also quite elongated.

Fact File

Name	Type	Visibility	Distance
M65	Galaxy	D	24 million light years
M66	Galaxy	D	33 million light years

Point Score

●●●●●●●●●●	Date seen	Points
Finding M65 and M66		5 each
Score		

Ursa Major, the Great Bear
Mizar

M82
M81
Alcor
Mizar
Alkaid
Alioth
Megrez
Dubhe
Merak
Phecda
M97
URSA MAJOR

The seven bright stars of Ursa Major probably make it the best-known constellation of all.

Where to look

Ursa Major is always visible from most of the northern hemisphere, as it is close to the sky's north pole. It's always somewhere to be found when you look in that direction, but it's most noticeable on fall evenings, when it hangs low on the northern horizon and people say "There's the Big Dipper." The main seven stars of Ursa Major make up what Americans call the Big Dipper, people in Britain call the Plough, and the French call "*la Casserole*" (saucepan).

But at other times of the year it has rotated around the Pole and it's not so obvious. In winter it is higher up in the northeastern sky, standing on its handle, in spring it's overhead, and in summer it is high up in the northwest, this time with its handle pointed upward.

These days ploughs (or plows) don't look anything like the pattern of the seven

Point Score		
●●●●●●●●●	**Date seen**	**Points**
Finding the Big Dipper		1
Seeing Alcor with the naked eye		1
Seeing Mizar as a double star		2
Score		

See also: **M81 and M82; M97.**

stars, but the original ones really did. In the same way, American homes would always have had a long-handled dipper, used to fetch water from the stream, but again things have changed.

Legends

Many peoples all over the world have thought of this constellation as a bear, though today we probably see it as a

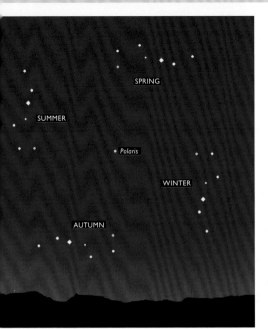

SPRING

SUMMER

Polaris

WINTER

AUTUMN

◀ How the Big Dipper appears in the evening at each season. Polaris is the Pole Star, in Ursa Minor, and is always due north at the same place in the sky.

with Callisto as a result of which she had a son, Arcas.

Years later, Arcas was out hunting and was about to kill a bear, little suspecting that it was in fact his mother. To prevent this, Zeus grabbed the bear by the tail and flung her into the heavens. She was now safe, but at a cost of being a bear with a very stretched tail!

What you'll see

The seven main stars of Ursa Major are only a part of the constellation. Its head and paws extend over quite an area of sky, and it is actually the third largest of the constellations.

Look at the middle star of the bear's tail, or the saucepan's handle if you prefer. This star is called Mizar, and there is another star, Alcor, quite close to it. Most people can see this star fairly easily, as it is quite bright (fourth magnitude) and is well separated from Mizar, but it has often been considered a test of eyesight. It is unusual for such a faint star to have a

saucepan. It lurks in the north, like bears (the Greek word for "bear" is *arktos*, from which we get *arctic*). But the Ancient Greeks needed a good story to go with this, so they linked the constellation with Callisto, a nymph or nature goddess. She was turned into a bear by Hera, the wife of Zeus, in revenge for Zeus having an affair

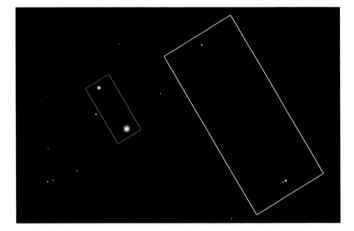

▶ Binoculars show Mizar and Alcor with another, fainter star nearby. Through a telescope (shown in the inset) you can see that Mizar is also a double star.

separate name, but Alcor draws attention to itself, being so close to a star in such a famous constellation. The Arabs thought of Mizar and Alcor as the "horse and rider."

Look at the stars through a telescope and you'll see that Mizar is also a double star, though the two are quite close together. If you like a challenge you could try seeing the two separately using binoculars, though you will probably need at least 10× binoculars to do so. There is another, fainter star making a triangle with Mizar and Alcor.

The list of distances of the main stars of Ursa Major shows something interesting

▼ In spring, look for the Big Dipper high overhead.

Brain Box

The seven bright stars of Ursa Major appear on the signs of many pubs in Britain called either "The Seven Stars" or "The Plough." The Seven Star Crags in China are also said to resemble the shape of Ursa Major.

– most of them are about 80 light years away from us. This is no coincidence. They also share the same movement through space, showing that they were all born in the same nebula, something like 500 million years ago. So you could think of the Big Dipper as being the largest star cluster in the sky!

Fact File

Name   Ursa Major, the Great Bear
Area   1,280 square degrees

Objects		Magnitude	Distance
Dubhe	Alpha Ursae Majoris	1.8	123 light years
Merak	Beta Ursae Majoris	2.3	80 light years
Phecda	Gamma Ursae Majoris	2.4	83 light years
Megrez	Delta Ursae Majoris	3.3	80 light years
Alioth	Epsilon Ursae Majoris	1.8	82 light years
Mizar	Zeta Ursae Majoris	2.0	86 light years
Alkaid	Eta Ursae Majoris	1.9	104 light years
Alcor	80 Ursae Majoris	4.0	82 light years

51

M81 and M82

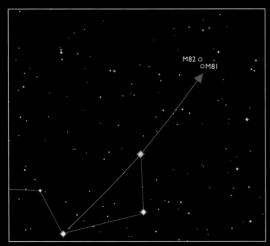

M82 ○
 ○M81

These two galaxies are a favorite among observers because of their interesting shapes.

Where to look

When Ursa Major – the Big Dipper – is high in the sky in spring, this is the time to look for these two galaxies. All you need to do is draw an imaginary line from Phecda through Dubhe and continue it as far again. This brings you to a pair of stars, of fourth and fifth magnitude. Come back toward the Big Dipper slightly, and you should spot these two galaxies.

There is another group of stars close to this line, but it is a triangle. Ignore this and try to find the pair that you want. The correct two stars almost point to M81, the brighter of the two galaxies.

Fact File

Name	Type	Visibility	Distance
M81	Galaxy	C	12.3 million light years
M82	Galaxy	C	12.7 million light years

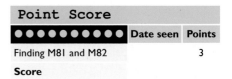

Point Score

Point Score	Date seen	Points
Finding M81 and M82		3
Score		

What you'll see

With 10 × 50 binoculars you can see the two galaxies looking like tiny fuzzy spots. They are not bright, and you need a reasonably dark sky to see them, so if your sky is a bit light polluted wait until Ursa Major is really high up and the sky is as good as you can get. Even with binoculars you can see that they are have some shape to them, and are not the usual circular blur that many deep-sky objects show.

A telescope gives a better view. M81 is larger and oval, while M82 is really quite thin and cigar-shaped. Take a good look and do a drawing if you can. Photographs show that M81 has lovely spiral arms. A galaxy

Those M numbers

Many bright deep-sky objects are listed in the *Messier Catalog* and have M-numbers. That's Messier in French, pronounced "Messi-ay," and not messier as your room gets when you don't tidy it!

Charles Messier was an 18th-century French comet hunter who made a list of troublesome objects that looked like comets as seen with his 50 mm refractor from Paris. He also added some obvious non-comets, such as the Pleiades, for completeness. There are 110 Messier objects in all. So all the objects with M numbers can be seen from the latitude of Paris with a small telescope – but that was in the days before light pollution!

▼ These are two very different galaxies, and they are bright enough that you don't need a big telescope to see them.

like this, with really obvious arms, is often referred to as a "grand design spiral." But M82 is quite different and seems to be a spiral galaxy seen almost edge-on, but with a tremendous burst of star formation occurring near its center.

▼ The author used 12 × 45 binoculars to make this sketch of M82 (left) and M81. If I can do it, you can!

The Owl Nebula, M97

This planetary nebula is a challenge for small telescopes in average conditions.

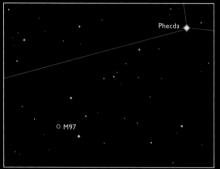

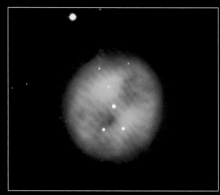

▲ A photograph through a large telescope shows why M97 gets its nickname of the Owl Nebula.

Fact File

Name	Owl Nebula, M97		
Type		Visibility	Distance
Planetary nebula		E	2,000 light years

Where to look

Start from Merak, which is at the bottom right of the Big Dipper's saucepan. With the finder, look to its left, along the line to Phecda (the bottom left star in the bowl), and if you have an optical finder you will see two stars at right angles to the line, with a third fainter star making a triangle with them. This is where to look for the Owl Nebula, which turns the triangle into a slightly distorted kite-shape.

What you'll see

The Owl Nebula needs a low magnification of about 30 to 50, as it is one of the larger planetary nebulae, though still less than a tenth of the diameter of the Moon. It shows as a faint ghostly disk, and is easily wiped out by a bit of light pollution so it's best to wait for the very clearest night with Ursa Major overhead in spring. This is an object that is helped by a light-pollution filter or an O III filter, which helps to make it stand out from the background, but you might end up wondering whether the expense was worth it!

The nebula gets its name from two darker patches within the disk, supposedly looking like the eyes of an owl. You need fairly good conditions and a telescope larger than about 200 mm aperture to see these.

Point Score

●●●●●●●●●●●	Date seen	Points
Finding M97		5
Score		

Virgo, the Virgin
Virgo Cluster

The galaxies in the great Virgo Cluster are among the most remote objects that you can see with a small telescope or even binoculars.

▲ The main part of the Virgo Cluster, with giant galaxies M87 (far left), M86 and M84.

Where to look

Virgo is in the sky during the spring. The first step is to work out which part of the sky you need to find. The time to look is when the Big Dipper is right overhead. Follow the curve of its handle downward and you come to a very bright yellowish star, Arcturus, in mid sky. Keep going in a curve and closer to the horizon is a blue-white star, Spica. This is the main star in Virgo. The other main stars of Virgo form a sort of Y-shape upward and to the west (right) of Spica.

To find the Virgo Cluster of galaxies, next you need to find **Leo**, which is in mid sky below the Big Dipper. It really does look like a join-the-dots picture of a crouching lion. The star that marks its tail is called Denebola, and this is your starting point to find some of the members of the Virgo Cluster.

From Denebola find a fifth-magnitude star about 7° eastward, 6 Comae. Next,

Point Score

●●●●●●●●●●	Date seen	Points
Seeing any galaxy in the Virgo Cluster		6 each
Score		

move a similar distance southeastward to find another star of similar brightness with a fainter star just above it, Rho Virginis. Almost exactly halfway between 6 Comae and Rho Virginis lie some of the brightest members of the Virgo Cluster. Use the chart overleaf to pick out some of the fainter stars to help you spot M84, M86 and M87.

Legends

This is one of those constellations whose legends have become mixed up over the

6 Comae

M86 ○ ○ M84

M87 ○

Rho Virginis

Fact File

Name	Magnitude	Type	Visibility	Distance
Spica Alpha Virginis	1.0	–	–	250 light years
M84	–	Galaxy	E	60 million light years
M86	–	Galaxy	E	55 million light years
M87	–	Galaxy	E	56 million light years

The first thing to say is that these galaxies really are small, and the second thing is that you do need a good, clear, dark sky. If your sky is so poor that you can't see any stars in between 6 and Rho with your binoculars or telescope, don't bother looking for galaxies but try again when you have darker or clearer skies. The galaxies are faint and fuzzy, and look gray rather than white. Photos always make them look bright in the middle, but they are actually more ghostly than that even in the middle. They are also quite small, and in each case they are brighter in the center than at the outside.

years. She appears in spring, and the star Spica is usually shown on old depictions as an ear of wheat, representing the fertility of the soil. She is also seen as an innocent young goddess who lived in an early golden era before wars, illness and greed, and who fled to the heavens when the world changed for the worse.

And she is also the goddess of justice, whose scales (used to weigh one argument against another) are represented by the nearby constellation of Libra. Unfortunately, Libra has few objects of astronomical interest, so gets no further mention in this book!

What you'll see

Being far from the Milky Way, there are no glorious star clusters or nebulae here. Instead, we are looking out to the Universe beyond, and to the huge Virgo Cluster of galaxies.

The easiest ones to see are M84 and M86, as these are quite bright and are close together, with M87 slightly brighter and larger to their southwest. But don't expect to see lovely spiral arms. The brightest galaxies are generally elliptical rather than spiral, and the sad fact is that all you can see with a small telescope is the inner part of the galaxy anyway, so any spiral arms are too faint to be seen.

If conditions are really good you might go on to spot others, but finding your first galaxies in the Virgo Cluster is a great thrill. This is a huge cluster of galaxies, with well over 1,000 members in all. By studying stars in the Virgo galaxies, particularly using the Hubble Space Telescope, astronomers have been able to work out its distance – around 55 million light years. This has helped them to estimate the size of the Universe as a whole.

Arcturus

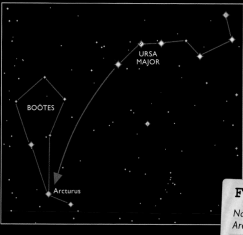

Boötes (pronounced "Bo-oh-teez," not Boots!) is mostly famous for brilliant Arcturus, which dominates spring skies.

Fact File

Name	Boötes, the Herdsman
Area	907 square degrees

Object	Magnitude	Distance
Arcturus Alpha Boötis	0.0	37 light years

Where to look

Find Arcturus and you have found Boötes. Arcturus is unmistakable in the spring and summer skies. It is a bright and clearly orangey star fairly high up. Follow the curve of the handle of the Big Dipper and you come to Arcturus.

Legends

Boötes is a herdsman of the Great Bear itself, whom legends said was the goddess Callisto, who had been turned into a bear in another Greek myth. The herdsman was Arcas, the son of Zeus and Callisto. One day Arcas came across the bear, who, as any mother would, called out to her son. But Arcas only heard a growl, and gave chase. Callisto fled to Zeus, who avoided a tragedy by placing Callisto in the sky as the Great Bear (see page 128) and Arcas as Boötes, her herdsman, following her through the sky forever.

What you'll see

Arcturus is a good signpost star, being so bright and noticeable – it's the fourth brightest star in the sky. It is a red giant star, which gives it its orange color, and shows what the Sun will look like in about 5 billion years, when it too becomes a red giant. Arcturus is only slightly more massive than the Sun, and at one time was probably very similar, but now it has swelled up to become 25 times bigger and 170 times brighter.

Point Score

⊙⊙⊙⊙⊙⊙⊙⊙⊙⊙	Date seen	Points
Finding Boötes		1
Score		

135

55

Coma Berenices, Berenice's Hair
Coma Star Cluster 👁 🔭

The only constellation named after a real person, **Coma** contains a large star cluster which is a test of a dark sky.

Star map showing URSA MAJOR, Arcturus, Beta, Alpha, Gamma, COMA BERENICES, LEO, and Denebola.

Point Score

	Date seen	Points
Finding Alpha, Beta and Gamma Comae Berenices		I each
Finding the Coma Star Cluster		2
Score		

Where to look

When Arcturus is high in the sky, in spring and summer, you can look for the rather faint stars of Coma Berenices, often simply called Coma. It is marked by three fourth-magnitude stars, which are arranged in a triangle like a set square. The bottom one, Alpha Comae Berenices, lies between Arcturus and Denebola in **Leo**, though rather closer to Arcturus. From there, look about half a stretched hand's width toward the Big Dipper and you will find another fourth-magnitude star, Beta Comae, then turn through a right angle toward the top of the Sickle of Leo and you will find Gamma Comae about the same distance away. Hanging down from Gamma is the Coma Star Cluster.

Legends

Berenice (pronounced "Berry-ny-see") was an actual queen of Egypt in the 3rd century BC. The story goes that she offered to cut off her flowing locks as an offering to the goddess Aphrodite if her husband was victorious in a battle. When

Brain Box
We think of the Sun as an average star, but there are very few similar nearby stars. Beta Comae is one of them, being slightly larger and brighter than the Sun. It is not known to have planets.

◄ Photographs bring out the colors in the stars, but the Coma Star Cluster is a pretty sight anyway.

Fact File

Name	Coma Berenices, Berenice's Hair	
Area	386 square degrees	
Objects	Magnitude	Distance
Alpha Comae	4.3	58 light years
Beta Comae	4.3	30 light years
Gamma Comae	4.4	167 light years

he returned safely having won, she duly cut off her hair and placed it in the temple of Aphrodite. But for some reason the hair went missing, and to soothe the queen the court astronomer pointed at the constellation, saying that her hair was now in the heavens. However, it did not become recognized as a constellation until the 16th century.

If only people were as gullible these days about what astronomers tell them!

What you'll see

In a dark country sky you can see the Coma Star Cluster with the naked eye. It is about 5° across, which is about the field of view of many binoculars – much larger than other clusters such as the **Pleiades** or the Beehive (**M44**). There are only a handful of stars visible with the naked eye, and some of those are not actually part of the cluster, in the general shape of a sprig of mistletoe with two branches. Gamma Comae itself is not actually a genuine cluster member, being rather closer, but it adds to the appearance.

Although you can only see a few individual stars even in a good, dark sky, the cluster gives the impression that there is more to be seen, particularly if you don't look directly at it (the technique of averted vision, see page 110). This is because the fainter stars do add to the overall brightness of the area, even though you can't see them individually. You can see them with binoculars, though, even from suburban areas. But try your telescope and you might be disappointed, as it probably has too high a magnification to show the cluster as a group surrounded by darker sky.

Incidentally, there is another Coma Cluster, a cluster of galaxies, which is too faint to be seen with small telescopes.

137

56

Canes Venatici, the Hunting Dogs
M3

Some of the best sights in the sky are within this little-known constellation of spring skies.

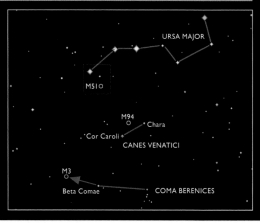

Where to look

The best time to look for Canes Venatici is when the Big Dipper is high overhead in spring. If you think of five of the stars of the Big Dipper as forming part of a circle, Canes Venatici lies at the center of that circle, with its main star, third-magnitude Cor Caroli, near the middle. There are just two brightish stars: Cor Caroli (Alpha) and the fainter Chara (Beta).

Legends

This isn't an ancient constellation – it was created in the 17th century by Polish astronomer Johannes Hevelius, who obviously felt that the herdsman Boötes needed a couple of dogs. But there is an interesting story about the naming of Cor Caroli, which is Latin for "heart of Charles." The Charles referred to is King Charles I of England, who was executed by republicans in 1649 following the English Civil War. The star is also claimed to have shone particularly brightly on the night of May 29, 1660, when his son, Charles II, was restored to the monarchy.

What you'll see

Apart from Cor Caroli and Chara, the stars of Canes Venatici are fifth magnitude or fainter and don't make up any special pattern. Cor Caroli itself is a double star,

Point Score

●●●●●●●●●●	Date seen	Points
Finding Cor Caroli and Chara		2
Finding M3		2
Score		

See also: **M94; M51.**

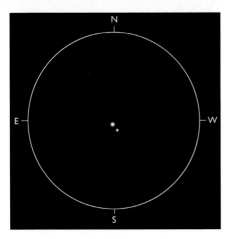

▲ A drawing of the double star Cor Caroli by Jeremy Perez, as viewed through a 150 mm reflector with a magnification of 240.

Coma Berenices. Begin at the top edge of the Coma Star Cluster and move due east to find the star Beta Comae. There are a few fainter stars along the line to keep you on track. Keep going half as far again and you come to a fuzzy star, which is M3. There is a finder chart on page 20.

This is one of the brightest globular clusters in the northern hemisphere, and when it is virtually overhead it is ideally placed. It is not as bright as M13, but is noticeably brighter than M92, both in **Hercules**. With a small telescope of around 100 mm aperture you can get the impression of individual stars – it has a sort of lumpy or grainy look to it, rather than having a smooth surface.

which you can see with a magnification of 20 or more with a small telescope. The fainter star is just slightly yellower than the brighter one, but some people see the two as being blue and orange. Take a look and decide for yourself.

The genitive version for Canes Venatici is even more complicated – Canum Venaticorum! So many people just use the three-letter abbreviation of CVn instead.

The brightest deep-sky object in Canes Venatici is the globular cluster M3. But the best way to find this is to start from the neighboring constellation of

▶ As globular clusters go, M3 is a good one so it's worth studying with as much magnification as you can muster.

Fact File

Name	Canes Venatici, the Hunting Dogs				
Area	465 square degrees				
Objects		Magnitude	Type	Visibility	Distance
Cor Caroli	Alpha Canum Venaticorum	2.8	–	–	115 light years
Chara	Beta Canum Venaticorum	4.3	–	–	27 light years
M3		–	Globular cluster	B	33,000 light years

57

The Whirlpool Galaxy, M51

The famous Whirlpool Galaxy is often featured in photos, and you may even pick it out with binoculars.

Where to look

Although M51 is in Canes Venatici, the easiest way to find it is to start with the Big Dipper. Wait until it's high in the sky, in spring or early summer, because M51 is faint and you need the darkest sky possible. Go to Alkaid, the end star in the handle of the Big Dipper (or saucepan, if you prefer).

Near Alkaid, and roughly at right angles to the handle, is a fifth-magnitude star, 24 CVn, which is just inside Canes Venatici. M51 makes a triangle with this star. Look at the finder chart carefully to see where it is in relation to a smaller triangle of three seventh-magnitude stars.

Point Score		
●●●●●●●●●	Date seen	Points
Finding M51		5
Score		

What you'll see

Don't expect to see a swirling mass of stars like you see in the photos! In binoculars you can see just a pair of tiny fuzzy spots if your skies are really dark. You will probably have difficulty seeing even these if your binoculars only magnify seven or eight times, but with 10× magnification they are just visible under the right conditions. In a telescope with even a low magnification of 30 or so you should see the two fuzzy spots more clearly, but they still don't look much like the photos. You need quite a large telescope and really dark skies to see the spiral arms. In perfect conditions a 200 mm telescope will just show them, but most of the time you'd need a telescope two or three times this size.

This is really what astronomers call an interacting pair of galaxies, because they

are so close together that the larger one distorts the smaller one. The smaller one, called NGC 5195, is being drawn in by the gravity of M51 itself and will probably merge with it over hundreds of millions of years.

Fact File

Name	Whirlpool Galaxy, M51	
	Visibility	Distance
Type		
Galaxy	D	26 million light years

▲ Mizar is at top right in this photo, and Alkaid to the left of center. The photo emphasizes the brightness of M51 but you can see that it is quite small in the sky. The view here is about 10° from top to bottom.

▶ The pink bits in the arms of the Whirlpool Galaxy that show up on photos are giant clouds of hydrogen gas like the Orion Nebula in our own Galaxy. There's another photo of the Whirlpool on page 18, with south at the top as seen through a telescope. This picture (right) has north at the top as seen through binoculars.

How much you see of this interesting galaxy depends on how good your skies are.

◄ This is what you have to imagine when you see M94 through your telescope: a photo taken using a 200 mm telescope.

Where to look

First find Cor Caroli and Chara, the two brightest stars in Canes Venatici. These stars are at the ends of a flat triangle with M94, so aim your telescope roughly halfway between the two stars and move a short distance in the direction of Alkaid, the star at the end of the handle of the Big Dipper. M94 is about 2° away, which is about two eyepiece fields of view. (Use the finder chart on page 138.)

Fact File

Name	M94	
Type	Visibility	Distance
Galaxy	D	16 million light years

What you'll see

You're looking for a fuzzy star, one that looks slightly out of the ordinary. The galaxy M94 is face-on to us, and its center is very condensed, so it looks very much like an ordinary star of eighth magnitude or so. You should spot it with a magnification as little as 30, but once you've found it you can increase the magnification a bit.

The darker your skies, the more it looks like a galaxy and the less like a star. Photos show it having numerous spiral arms and a fainter ring farther out.

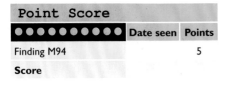

Point Score

●●●●●●●●●●	Date seen	Points
Finding M94		5
Score		

OBJECTS TO LOOK FOR IN SUMMER

Now for objects that you can see in the summer skies, though you may also find some of them in spring as well.

In summer it gets dark later, so the all-sky map on this page applies for August 1 at 10.00 pm, but you can recognize the same constellations a month or two on either side of that or later in the evening. In northern states and Canada it isn't fully dark at this time.

Start by looking high in the sky for the two bright stars Arcturus in Boötes and Vega in Lyra. Then down near the southern horizon is Antares in Scorpius.

The exact position of stars at the top and bottom of the map will vary depending on how far north or south you live.

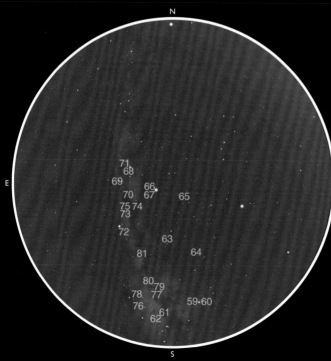

This map was created using Stellarium software, which you can download free from www.stellarium.org.

Scorpius, the Scorpion
Antares

Summer skies have a fierce scorpion lurking down in the southern sky, its red heart marked by the star Antares.

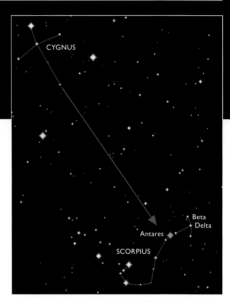

Where to look

Because Scorpius is quite far south, in the northern states and Canada it's only above the horizon for a short time. And because that's in summer, when it doesn't get dark there until late, spotting the stars of Scorpius means looking at just the right time.

You can start looking for it in the southeast very late on May nights, but it is best seen in the south on July evenings and by the end of September it is sinking into the twilight in the southwest about two hours after sunset. If you can find **Cygnus**, high overhead at this time of year, the head of the Swan points down to Scorpius close to the horizon.

The map here shows the whole of Scorpius, but you won't see the very bottom of it from northern areas.

Point Score ●●●●●●●●●●	Date seen	Points
Finding Scorpius		I
Identifying Antares, Beta and Delta		I each
Score		

Legends

To many ancient peoples, the very shape of Scorpius resembled a scorpion, with its curved line of stars being the upraised tail and sting, other stars for its claws and reddish Antares at its heart. The ancient Greeks saw it as the scorpion that killed Orion. The great hunter had boasted that he could kill any animal, but to teach him a rather severe lesson, rather than confront him with a mighty beast he was stung by the tiny scorpion.

Now that both Orion and the scorpion are among the stars, they are kept so far apart that they are never in the sky together. As Scorpius rises, Orion sets; and when Orion starts to appear, the wily scorpion slips below the opposite horizon.

To astronomers, the constellation is called Scorpius. It's only astrologers who call it Scorpio, so watch your language!

What you'll see

At the heart of Scorpius is the red giant star Antares, which has a fainter star on either side of it, marking the creature's body. This short line of stars points upward and to the right to a group of stars marking its claws. The star Delta Scorpii (Dschubba) almost doubled its brightness

---

Red giant stars

When a typical star nears the end of its life and starts to use up all its hydrogen fuel, it swells up to an enormous size. But its temperature goes down, which means that its light becomes much redder. The Sun won't do this for at least 4 billion years, so don't worry about it happening to us!

---

in 2000, and can be brighter than Beta Scorpii. Keep an eye on it.

Go the other way and you come to a curve of stars that dips down southward and then curves back upward to a group of stars marking the sting. But people in the northern states and Canada are at a disadvantage because this line of stars is too far south to be seen, so you will have to wait until you go farther south to see them.

▼ Scorpius skims the chimney tops for many people in northern areas.

Fact File

Name	Scorpius, the Scorpion		
Area	497 square degrees		
Objects		Magnitude	Distance
Antares	Alpha Scorpii	1.1	552 light years
Graffias	Beta Scorpii	2.5	530 light years
Dschubba	Delta Scorpii	2.3 to 1.6	490 light years

Scorpius contains one of the easiest globular clusters to find, as long as you have the right viewing conditions.

► M4 is due west of Antares, though at just over a degree away you will need to move your telescope to see it, even using a low-power eyepiece.

Where to look

If you can find Scorpius, you can look for the globular cluster M4, because it is right next to **Antares**, the red star at the heart of Scorpius.

Antares is flanked by two other stars, the right-hand one being Sigma Scorpii. Making a little triangle with Antares and Sigma is M4. This is quite easy to see with binoculars if you have good clear skies, but it's a challenge if Scorpius is low down and in the murk. Observers in northern areas will need to choose their night, because a bit of haze or light pollution will be enough

to wipe it out. But at least you can pinpoint its location quite easily, so looking for it should be a cinch. The photo here and the map on page 148 show where to look.

With typical binoculars you can get Antares and Sigma together in the same field of view, so M4 is midway between the two and a short distance below the line between them.

What you'll see

Although most globular clusters are quite far away, M4 is comparatively close and is large even in binoculars, as globular clusters go. Altogether, including the faint outer edges, which you won't see in poor skies,

Fact File

Name	M4		
Type		Visibility	Distance
Globular cluster		B	5,600 light years

Point Score

Point Score	Date seen	Points
●●●●●●●●●●		
Finding M4		3
Score		

it appears about the size of the Moon. It looks like a pale fuzzy blob, a bit brighter toward the center but nothing like as bright as it appears in most photos. With a telescope you can start to see that it's a ball of stars, and in a medium-sized telescope it can be spectacular on a good night.

▼ Imagine being on a planet in the middle of a globular cluster like M4 – the sky would be filled with bright stars. But astronomers believe that the stars are so close together – typically only the size of the Solar System between them – that planetary systems would be disrupted by the gravitational pull of neighboring stars.

Globular clusters

These are rather different from ordinary star clusters, which are also called open or loose clusters. Globular clusters, as well as being ball-shaped, are not in the spiral arms of the galaxy, like open clusters, but surround it in a sort of halo. They are generally much more distant than open clusters, but they contain many more stars. A big globular cluster, such as Omega Centauri, in the southern constellation of Centaurus, contains several million stars, though M4 probably has about 750,000. Even a large open cluster has only a few thousand stars in it, and most have fewer.

A bright star cluster that glows within the Milky Way – as long as you are in the right place to see it.

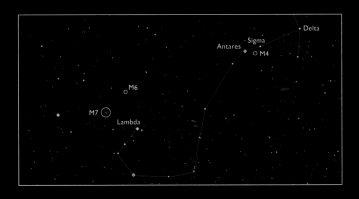

Fact File

Name Butterfly Cluster, M6

Type	Visibility	Distance
Open cluster	B–E depending on your location	1,600 light years

Where to look

Start with **Antares** in Scorpius and the two stars on either side of it. Follow these down to the lower left and you come to a fairly bright star, Lambda Scorpii, which marks the sting of the scorpion. Above this, and to the left a bit, is the cluster M6. If Lambda is at the bottom of the field of view of ordinary binoculars, M6 should be near the top.

These instructions are all very well if you are far enough south that you can actually see Lambda Scorpii, but from Canada it is pretty low even at the best of times and from north of Calgary it doesn't even get above the horizon! So you'll just have to guess its position, and scan with binoculars to find M6, which could require a lot of finding! You might need to wait for a trip to a more southerly location to see it.

What you'll see

From a good site, M6 is easily visible with the naked eye as a small hazy area in an otherwise darker region of the Milky Way. It's one of many such tantalizing patches that you see when you have a really good view of the Milky Way. Turn binoculars on it and you can see a good number of stars within the cluster, the brightest being about seventh magnitude. They cover about the same area of sky as the full Moon.

Look carefully and you'll see that the stars happen to lie in curved arcs, which resemble the wings of a butterfly. Many people refer to M6 as "the Butterfly Cluster."

For just 40 minutes in 1965, one of the brightest stars in M6 was seen to flare up to magnitude 2, though it seems to have behaved itself ever since.

Point Score

● ● ● ● ● ● ● ● ●	Date seen	Points
Finding M6 from a good site		2
Finding M6 from a bad site		5
Score		

This star cluster is a beautiful sight in binoculars – but it's at the limit of visibility for many observers.

▶ Both M6 and M7 are shown on this photo taken from New Zealand – but they are a challenge to observe from the northern US and Canada.

Point Score

	Date seen	Points
Finding M7 from a good site		2
Finding M7 from a bad site		5
Score		

Where to look

If you can find M6 in Scorpius, M7 is just a short distance away. First find M6, then look to its lower left about the width of a field of view of binoculars, and there is M7. From southerly latitudes this is easy enough, and in fact M7 is the more easily visible of the two clusters, but as with M6, it's a major challenge for northerly observers. You will need to wait for the clearest of nights and maybe travel to a more southerly dark site to have any hope of success. (Use the finder chart on page 148.)

What you'll see

Unlike M6, which is in a dark zone within the Milky Way, M7 looks to the naked eye like a brighter part of the Milky Way. It's much larger than M6 and from the right location is much more impressive, with a dozen or so stars easily visible with binoculars and many more if you have good skies and a telescope.

It covers more sky than twice the diameter of the Moon, which means that in larger telescopes and at higher magnifications it might not look as good as in binoculars unless you have a very expensive wide-angle eyepiece. Clusters usually look their best when there is a bit of darker sky around them.

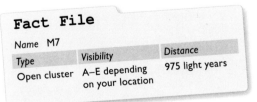

Fact File

Name	M7	
Type	Visibility	Distance
Open cluster	A–E depending on your location	975 light years

Ophiuchus, the Serpent-Bearer IC 4665

Despite being a large constellation and easy to observe, **Ophiuchus** is often overlooked because it has so few bright stars.

NGC and IC numbers

Many deep-sky objects are cataloged with NGC numbers, as these are from the *New General Catalogue*. This was compiled by J. L. E. Dreyer at Armagh Observatory in Ireland in 1888. He issued two supplements, the *Index Catalogues* (IC) a few years later. Together, they include over 13,000 objects. Most deep-sky objects visible with small- to medium-sized telescopes have either an NGC or an IC number in addition to any M numbers.

Where to look

You can find Ophiuchus from late evenings in May until early evenings in November. It's located below **Hercules**, though that isn't a particularly prominent constellation either. It's best visible when both Arcturus and Vega are in the sky together, being the two brightest stars visible in summer.

The easiest constellations to recognize are those with a nice, compact group of stars that form a recognizable pattern. However, the stars of Ophiuchus are anything but. The shape they make is nothing in particular, and it's spread out over a large area of sky. From a light-polluted area most of them are quite hard to see, so the whole thing takes some looking for. To find it for the first time, start at Vega and find Hercules by going toward Arcturus. Use the Keystone of Hercules as an arrow to point into the rather barren area below Hercules and you will spot a pair of stars fairly close together, called Yed Prior and Yed Posterior. These mark one edge of the pattern.

Follow the line of the two stars down and to the left and you'll encounter two

Point Score

●●●●●●●●●	Date seen	Points
Finding the main stars of Ophiuchus		2
Finding IC 4665		3
Score		

▲ The field of view from top to bottom of this photo of IC 4665 is about the same as 10 × 50 binoculars.

lower left. There is actually a constellation called Serpens in these areas, which is in two bits, Serpens Caput being the head and Serpens Cauda being the tail.

In Greek legends, Ophiuchus is Asclepius, god of healing. It may seem strange that a healer is linked with a snake, but the thinking was that as snakes can shed a skin and become apparently new again, maybe they have some clever healing powers. Today, the sign of a staff entwined with snakes is still a symbol of medicine.

evenly spaced stars, marking the bottom of the pattern. To find the top end, go back to the Keystone and look to its lower left, where you'll spot a brightish star, Rasalhague, the brightest star in Ophiuchus, with another star, Cebalrai, some way to its lower left. From there you can pick out the other stars, such as the little triangle to the left of Cebalrai.

Legends

On classical star maps, the ones with pictures of monsters and bits of people all over the place, Ophiuchus is shown with the head of a serpent at top right and its tail at

What you'll see

Having identified the main stars of Ophiuchus, go to Cebalrai and look just to its upper left to find a star cluster known as IC 4665. This is quite an easy cluster to find with binoculars even from fairly poor skies, so it's odd that it isn't in either of the main two catalogs of deep-sky objects, the *Messier Catalog* (see page 131) or the *New General Catalogue*. The reason is probably that while it's quite obvious in binoculars, its couple of dozen stars are quite widely scattered so they hardly show up as a cluster in a telescope.

Also look for

Within the central area of Ophiuchus are the globular clusters **M10 and M12**.

Fact File

Name	Ophiuchus, the Serpent-Bearer				
Area	948 square degrees				
Objects		Magnitude	Distance	Type	Visibility
Rasalhague	Alpha Ophiuchi	2.1	258 light years	–	–
Cebalrai	Beta Ophiuchi	2.8	82 light years	–	–
IC 4665		–	1,150 light years	Open cluster	C

151

These two globular clusters can be seen together in the same field of view of binoculars.

M12 o

M10
30 • o M10

Yed Prior
Yed Posterior

◄ The field of view of this photo is about 4°, so the two clusters should fit easily together in one view. But you'll need to star-hop if you are using a telescope.

Where to look

Your starting points for this voyage into the depths of Ophiuchus are the two stars Yed Prior and Yed Posterior, some way below the Keystone of Hercules. With these stars central in your binoculars or optical finder, look a couple of fields of view to their east (left) and you come across a fifth-magnitude star called 30 Ophiuchi. M10 is just to the right of this star, and M12 is above and to the right of that. Check the finder chart to make sure you have your angles right when star-hopping to these clusters.

Point Score

○ ○ ○ ○ ○ ○ ○ ○ ○	Date seen	Points
Finding M10		3
Finding M12		3
Score		

What you'll see

These globular clusters look quite similar to each other, and you can compare them easily because they are so close together. With small or low-magnification binoculars they look like fuzzy stars, but with a telescope you can start to see some individual stars. M10 has a slightly brighter center, while M12 is more spread out, and also has a few obvious brighter stars of tenth magnitude that are much closer than the cluster itself.

Fact File

Name	Type	Visibility	Distance
M10	Globular cluster	D	14,000 light years
M12	Globular cluster	D	16,000 light years

Hercules
M13 and M92

65

Famous for its globular clusters, this constellation is quite faint but fairly easy to spot in the summer sky.

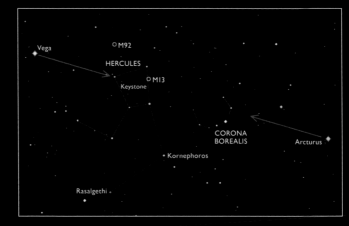

Where to look

Hercules is a feature of summer skies, but it's visible in the evenings from spring, when it rises late at night in the east, through to autumn when it sets in the west in early evening. It lies just to the west of the bright star Vega, which is high up in the sky in summer. It's on a line between Vega and Arcturus, another spring and summer star, which is the other very bright star high in the sky in summer.

From Vega, trace a line toward Arcturus, then Hercules is a third of

Soundbite
Hercules is pronounced "Her-ku-lees," or "Herra-klees" to the Greeks.

the way along that line. Its most obvious pattern is called the Keystone, because it is shaped like the wedge-shaped keystone in the middle of an arch. Most of the constellation's other stars dangle from the bottom ends of the Keystone.

Legends

Hercules, or Heracles as he was known to the Greeks, was the greatest hero of them all. He had a good start in this because he was immortal – basically the mythical

Point Score		
●●●●●●●●●●●	Date seen	Points
Finding Hercules		1
Finding M13		2
Finding M92		3
Score		

Brain Box

Globular clusters such as M13 and M92 contain some of the oldest stars you will see with your telescope. They probably formed nearly 13 billion years ago, a billion years or so after the start of the Universe.

◄▼ M13 (left) is larger than M92 (below) but the latter is more condensed in the center.

Fact File

Name Hercules (name of legendary character)
Area 1,225 square degrees

Objects		Magnitude	Distance	Type	Visibility
Rasalgethi	Alpha Herculis	3.1 (variable)	359 light years	–	–
Kornephoros	Beta Herculis	2.8	139 light years	–	–
Great Hercules Cluster, M13		–	23,000 light years	Globular cluster	B
M92		–	27,000 light years	Globular cluster	D

version of Superman. But instead of saving people falling from skyscrapers he was set ten tasks, such as killing a troublesome lion and cleaning out the Augean stables, which were particularly unpleasant. These were the Ten Labors of Hercules, and people still talk of "a herculean task."

What you'll see

The most famous feature of Hercules is the globular cluster M13. In old books it's often referred to as the Great Hercules Cluster, so you might think it is something obvious. But even with binoculars it's hardly visible, and then only from a fairly dark site.

The place to look is the right-hand side of the Keystone. Look a third of the way down and you should see a slightly fuzzy star, which is the cluster M13 that all the fuss is about. The higher the magnification of the binoculars, the better it appears, so with small 7× or 8× binoculars it might not look like anything at all. With higher magnifications, such as 10× or 15×, you can see that it is definitely fuzzy. Through a telescope it becomes more obvious, and if you can push the magnification up to around 50 you can start to see that the fuzziness is caused by lots of fainter stars, all in a ball-shaped (globular) cluster.

The larger your telescope and the better your viewing conditions, the more stars you are likely to see, but even from a town you can really start to see why this was called the Great Cluster. It is one of the brightest globular clusters (see page 147), though by no means the brightest. But from the northern hemisphere it can get very high in the sky, which makes it easier to see and more spectacular than some others such as M22 in Sagittarius, which is quite low down.

Also look for

Having found M13, now you can go on to another globular in Hercules, M92. To find this, make a triangle with equal sides above the top edge of the Keystone. You'll find M92 just to the left of this point. It's not quite as bright and is a bit more condensed than M13, so small binoculars may only show it as a star. There is a fifth-magnitude star to the south, but that is closer to the Keystone than an equal-sided triangle shows.

To the west of Hercules lies a pretty semicircle of stars, which is a constellation in its own right: Corona Borealis, the Northern Crown. The stars are not bright, but it is quite easy to spot if the light pollution is not too disastrous.

66

Lyra, the Lyre
Epsilon Lyrae

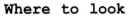

Epsilon Lyrae

Vega

Beta Lyrae

Gamma Lyrae
M57

You can't miss the brilliant summer star Vega, but its neighbor Epsilon Lyrae is a great test of your eyesight.

Where to look

Vega, the main star in Lyra, is one of the brightest stars in the sky, and is virtually overhead on summer nights. It is noticeably blue-white, so there's no mistaking it. You can't even confuse it with a planet, because these are lower in the sky and are never overhead in summer from the northern hemisphere.

Lyra, the Lyre, is one of the few ancient constellations that doesn't represent something alive and is the sky's only musical instrument. Unless you count Triangulum, which isn't meant to be a musical triangle anyway.

Legends

This was the lyre of the most famous legendary musician, Orpheus. His playing was so sweet that it could achieve miracles, including persuading Hades, the god of the Underworld, to release Orpheus's wife, Eurydice, from imprisonment there. The only condition was that Orpheus should not look round to check that Eurydice was still following. Orpheus did so until the very last

Fact File

Name	Lyra, the Lyre			
Area	286 square degrees			
Objects		Magnitude	Distance	Separation
Vega	Alpha Lyrae	0.0	25 light years	–
	Epsilon Lyrae	4.7, 4.6	162 light years	3½ arc minutes

moment as he emerged into the light, only to lose Eurydice forever by breaking the condition. No wonder they are called Greek Tragedies!

What you'll see

Apart from brilliant Vega, which is magnitude 0, the stars of Lyra are third magnitude or fainter. There is a little parallelogram of stars below Vega which is very recognizable, and just above this, to the east of Vega itself, is another star, Epsilon.

At first glance Epsilon is just another star, but with good eyesight you can see that it is actually not one star but two of similar brightness, very close together. These are thought to be a genuine double or binary star – that is, one where the two stars are actually close, and rotate around each other. But don't expect to see a change if you look night after night – they must take hundreds of thousands of years to complete an orbit. If you can't see the

Point Score

●●●●●●●●●●	Date seen	Points
Finding Lyra		1
Finding Epsilon Lyrae		1
Seeing Epsilon as two stars		2
Seeing each star as a double		3
Score		

See also: **M57.**

two stars separately by eye, binoculars will definitely show the pair.

But look through a telescope of 70 mm aperture or larger and you'll see that each of the individual stars is a double star in its own right, so Epsilon Lyrae is known as the "Double Double." One pair is at right angles to the other. In this case, the rotation times of the pairs are better known, one taking around 700 years and the other about 1,800 years. Even so, you're unlikely to see any change, even over a lifetime.

Soundbite

Euridice is pronounced "You-riddi-see."

▶ You need a magnification of about 200 to see the stars that make up Epsilon Lyrae as clearly as this. The star is a good test of a small telescope and the seeing – on bad nights the star images can be quite bloated.

67

The Ring Nebula, M57

A smoke ring in the sky – or a celestial donut? See for yourself and decide!

▶ This photo shows how small the Ring Nebula is compared with the distance between Beta (top right) and Gamma Lyrae (bottom left).

Where to look

When Lyra is high in the sky in summer or autumn you can look for the Ring Nebula, one of the showpieces of the sky. It is what's called a planetary nebula, and is the easiest to find of these amazing objects.

Find Vega, and then find the bottom two stars of the parallelogram below Vega: Beta and Gamma Lyrae. These are fairly far apart when seen through a telescope, so you'll need to move the telescope to get from one to the other. Use a magnification of around 50, because your target is actually quite small compared with many other deep-sky objects.

The Ring Nebula lies roughly midway between the two stars. You'll know it when you see it, as it is a faint ring rather than a star. (Use the finder chart on page 156.)

What you'll see

You can tell from its name that this is a ring, and that's just what you see. It looks like a tiny, ghostly ring, though with very small telescopes you might not see the central hole. You need a magnification of about 100 to show it well, and looking like a donut or smoke ring. Though the photos show it looking pretty with colors, these don't show up to the eye, though they are genuine colors caused by the gases in the nebula.

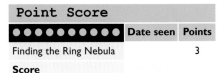

Point Score

	Date seen	Points
●●●●●●●●●●		
Finding the Ring Nebula		3
Score		

Fact File

Name	Ring Nebula, M57		
Type		Visibility	Distance
Planetary nebula		C	2,300 light years

It isn't circular, but elliptical – like a squashed circle. Photos show a star at the center, but you need a pretty large telescope to see this. This star was the cause of the nebula, as it threw off a shell of gas about 1,500 years ago. The shell extends all around the star, like a bubble, but we see it as a ring because the outer parts have denser gas than the center.

▼ This Hubble Space Telescope view shows the central 15th-magnitude white dwarf star that expelled shells of gas. The colors are due to different gases.

About planetary nebulae

Planetary nebulae are so-called because the first ones to be found were circular and about the same size as a planet as seen through a telescope. They are a stage in the life of a medium-sized star, when it has exhausted much of the hydrogen that keeps it shining normally. The star undergoes a drastic internal reorganization and throws off shells of gas, which expand around it. These are only temporary, but even so they last for thousands of years, and they don't change their appearance over a human lifetime.

Normally, stars are too small to be seen as anything other than points of light, but these shells of gas can easily be a light year in diameter, which makes them visible even with small telescopes. Many planetary nebulae are not circular but are oval, or are shaped like butterflies or even squares.

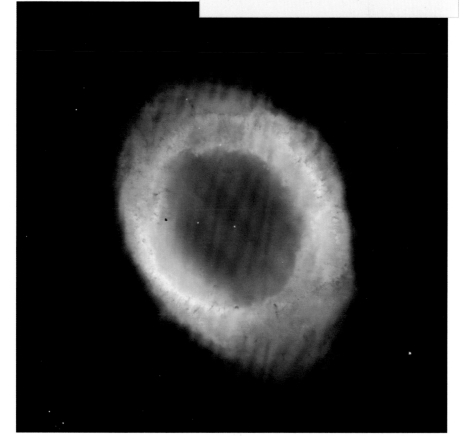

68

Cygnus, the Swan
61 Cygni
and Albireo

Cygnus is a key constellation of the summer skies, lying in the middle of the Milky Way.

Fact File

		Magnitude	Distance
Name	Cygnus, the Swan		
Area	804 square degrees		
Objects			
Deneb	Alpha Cygni	1.3	1,400 light years
Albireo	Beta Cygni	2.9	400 light years
	61 Cygni	4.8	11.38 light years

Where to look

Cygnus is overhead from North America in summer, but it is actually visible from late April evenings rising in the northeast, until early February when it sets in the northwest. Its basic cross-shape is also referred to as the Northern Cross, and you first see it lying on its side when it rises in the northeast in spring, with the main part of the cross parallel to the horizon. In August and September it is overhead, then as winter draws in it starts to sink into the northwest with the cross vertical.

It lies to the east of the very bright star Vega, which is also overhead in summer. So find blue-white Vega and look to its east, and you'll see the first-magnitude Deneb, which marks the top of the cross, or the tail of the swan which is flying along the Milky Way.

Legends

There is a very saucy tale about how the king of the gods, Zeus, was pursuing a goddess named Nemesis. To escape his unwelcome attentions she tried changing her form into increasingly swift creatures, ending up rather unwisely as a goose so that she could fly away. But Zeus promptly changed into the swan that we now see in the sky and caught up with her. The result of her encounter was that the goose Nemesis laid an egg, which eventually hatched into Helen of Troy. Today, the goose is nowhere to be seen in the sky, though there is a fox nearby, Vulpecula….

What you'll see

Cygnus works well as either a swan or a cross. The Northern Cross is usually just the five brightest stars. The swan figure is made from Deneb as the tail, with the arms of the cross as the two wings, though there are further stars beyond those of the cross-shape, and the star Albireo as the head. The swan flies directly along the line of the Milky Way, and is a useful guide to its path in light-polluted areas where it is not visible.

Two stars in Cygnus show you how much stars differ in brightness. Deneb is the brightest star in the constellation, but it is also very distant. Some books say that it is more than 3,000 light years away, though the most recent estimates put it at about 1,400 light years. Even so, it is easily the most distant bright star. Not far from it in the sky is 61 Cygni, at magnitude 4.8, and quite hard to spot unless you have a good sky. This is a close star, at only 11.4 light years. Just looking at these two in the sky together and knowing their distances gives you an idea of how much brighter Deneb is in reality.

Soundbite
"Al-birr-a-o" or "Al-bye-reo" are two ways to pronounce Albireo.

▶ The contrasting yellow and blue colors of Albireo make it probably the most viewed double star in the sky.

Point Score

Point Score		
●●●●●●●●●	**Date seen**	**Points**
Finding Cygnus		1
Finding 61 Cygni		2
Seeing Albireo as a double star		2
Score		

See also: **Great Rift; Veil Nebula; North America Nebula.**

Albireo, at the head of the swan, is one of the gems of the sky. It is a double star in which the brighter star is yellow and the fainter one bluish. The color contrast is easy to see, and the two stars are far enough apart that even binoculars will show them as two stars. But you need a magnification of at least 20 to see them clearly enough to make out their colors. How long the two stars take to rotate around each other is unknown, but it could be as long as 100,000 years.

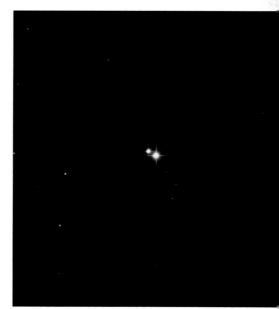

69

The Veil Nebula

One of the most beautiful objects in the sky as seen in photos, the **Veil Nebula** is a real challenge to find unless you have a filter.

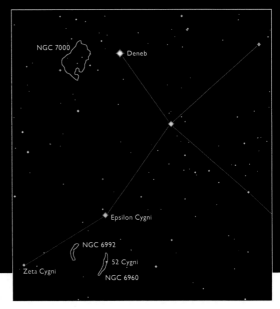

Where to look

Start by finding Cygnus, then pick out the star Epsilon Cygni, which is the left-hand star of the Northern Cross. Near this is a fourth-magnitude star called 52 Cygni. The brightest part of the Veil Nebula surrounds this star, extending on either side of it. There is another part, between 52 Cygni and Zeta Cygni, which some say is brighter but it is harder to find.

What you'll see

Probably nothing! If you have a telescope of aperture 130 mm or larger and a very dark sky, you might pick it out using a low-magnification eyepiece as a faint band of light extending either side of the star. Sometimes, dust or mist on the eyepiece can look similar, so prove the point by putting the star outside the field of view and moving the telescope a bit from side to side. Can you still see the band of light? If so, you are seeing the nebula.

This is an object best suited to telescopes with a low *f*-number. People with catadioptric telescopes that are *f*/10 or *f*/14 have trouble because the nebula usually fills the lowest magnification eyepiece and they don't notice it. You really need darker sky on either side to be able to spot it.

The Veil Nebula is the remnant of a massive star that exploded thousands of years ago in a supernova explosion. A supernova occurs when a star much

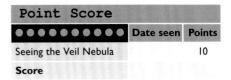

Point Score			Date seen	Points
●●●●●●●●●				
Seeing the Veil Nebula				10
Score				

more massive than the Sun runs out of the gas that causes it to shine, and for a short time becomes very brilliant.

▼ This section of the Veil Nebula, near to 52 Cygni (the bright star in this view), is also known as the Witch's Broom Nebula.

The O III filter

If you are really keen and can afford it, an O III filter (that's an "O" with a Roman numeral "III," and pronounced "O-three") helps a great deal. This cuts out all the light except for that from the oxygen gas in the nebula, so it gives a very dark view and turns the stars green. It screws into the back of the eyepiece and costs about as much as a cheap digital camera. A lot of money just to see a faint object, though it does help with finding some other nebulae as well, particularly planetary nebulae. But you need to use it with telescopes larger than about 100 mm or the view is too dark.

Fact File

Name	Veil Nebula, NGC 6960			
Type		Visibility	Distance	Age
Supernova remnant		E	About 1,470 light years	5,000–8,000 years

70

The Great Rift

The Great Rift is a dark nebula – that is, a cloud that doesn't shine at all. So how can you see it?

◀ This wide-angle view shows the Great Rift clearly. Evenings in August or September are best for viewing it.

Fact File

Name	The Great Rift		
Type	Visibility	Distance	Mass
Dark nebula	B	300 light years and beyond	About a million times the mass of the Sun

Where to look

You need to be able to find Cygnus, and you need a fairly clear and dark night when Cygnus is pretty well overhead. This is not something that you can see from the middle of a large city, but if you can get out into the country so that you can see the Milky Way, it should become obvious even if the Milky Way itself is not all that bright.

What you'll see

The Great Rift looks like a wide, dark band right down the middle of the Milky Way, starting just to the left of Deneb and carrying on down to the right of Altair, before finally getting lost somewhere in Ophiuchus. It is the largest dark nebula in the sky, and is made of the gas and dust from which stars will eventually form. It can only be seen silhouetted against the background of the Milky Way. Use the photo here to help you find it.

Other spiral galaxies that are like the Milky Way have similar dark lanes in them. The material in these lanes is not the same as the dark matter that you may have heard of, though. That is something completely different – it never shows up at all, and no one really yet knows what it consists of. All we know is that there is something between the stars other than the gas and dust (such as the Great Rift) and black holes and other things, which is having an effect on everything else through the pull of gravity.

Point Score

⬤⬤⬤⬤⬤⬤⬤⬤⬤⬤	Date seen	Points
Seeing the Great Rift		3
Score		

It's over four times as large as the full Moon, yet the North America Nebula is very hard to find even with binoculars – unless you know the tricks.

▶ The shape of the North America Nebula is defined by dark lanes as much as by bright areas. There are other wisps of glowing gas in the area, but the North America Nebula is the easiest to see visually.

Where to look

First find the bright star Deneb in Cygnus, then look just to the left of it. It shows up really clearly on photos, looking just like a map of North America, but seeing it with the naked eye is tricky. In photos, this nebula glows pink, but even the best telescope or binoculars will not show it as anything other than a gray haze.

What you'll see

The reason why people can't see the North America Nebula straight away – apart from having poor skies – is that they expect to see something sharp and clear and colorful like in the photos. But lower your expectations and you will spot it

Point Score		
●●●●●●●●●●●	Date seen	Points
Seeing the North America Nebula		7
Score		

under the right conditions, especially if you follow the tips given below. It appears just as a slightly brighter area of the Milky Way. (Use the finder chart on page 162.)

If you can't see the Milky Way fairly easily in Cygnus, it's not worth looking for the North America Nebula. But if you can, turn your binoculars on Deneb. The trick is to look for a dark lane running eastwards from Deneb. This forms the East Coast. Then look for a less obvious dark lane at an angle farther east in the sky, which forms the West Coast.

Fact File

Name North America Nebula, NGC 7000

Type	Visibility	Distance
Gaseous nebula	C	About 1,600 light years

Aquila, the Eagle
Eta Aquilae

This celestial bird is a prominent feature of the summer Milky Way and is a useful signpost.

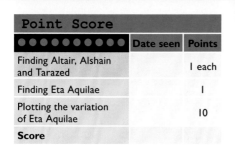

Point Score	Date seen	Points
Finding Altair, Alshain and Tarazed		I each
Finding Eta Aquilae		I
Plotting the variation of Eta Aquilae		10
Score		

Where to look

The most obvious feature of Aquila is its main star, Altair. This is flanked by two fainter stars, Alshain and Tarazed. Together they make an easy group to spot at any time from late evenings in May, when they rise over in the east, until the end of the year or even January, when they set in the west just after sunset. In late summer Altair is well up there in mid sky, forming part of the Summer Triangle whose other members are Vega in **Lyra** and Deneb in **Cygnus**.

Legends

Altair was thought of as the Eagle Star by many early peoples, and is said to be the eagle that stole Ganymede, son of the King of Troy, to become a servant to the gods. But in an alternative scenario from Korea and Japan, Altair and Vega are two lovers who have become separated by the river of the Milky Way. They can only meet once a year with the help of magpies that form a bridge across the Milky Way. The Japanese festival of Tanabata celebrates this event, though if it rains on Tanabata the two lovers cannot meet and must wait another year.

What you'll see

If you could see the sky in 3D, you'd spot straight away that Altair is one of the nearest bright stars to us. Most of the other stars are very distant, but Altair, at 17 light years away, would stand out, along with Vega at 25 light years. There are many other fainter ones that are also close, but these two are our local searchlights. In the winter sky, Sirius and Procyon are the nearby bright sparks at 8.6 and 11.4 light years, respectively. Nearest of all is Alpha Centauri, in the southern hemisphere, at just over 4 light years.

Despite being in the Milky Way, Aquila contains no bright nebulae or star clusters, though its masses of stars are lovely to scan with binoculars on a good, clear and dark night. Its main feature of interest is the star Eta Aquilae, a star whose brightness drops by more than half regularly every week (actually every 7.2 days). This is a type of pulsating star known as a Cepheid variable, and you can plot its slow variations for yourself by comparing it with other nearby stars. It's a very good star for learning the technique of measuring star brightnesses with the naked eye alone.

► Comparison stars for Eta Aquilae. Make notes of which star it is brighter or fainter than, then check their magnitudes afterward.

Fact File

Name	Aquila, the Eagle		
Objects		Magnitude	Distance
Altair	Alpha Aquilae	0.8	17 light years
Alshain	Beta Aquilae	3.7	45 light years
Tarazed	Gamma Aquilae	2.7	390 light years
	Eta Aquilae	3.5–4.4	1,380 light years

Use the chart here to decide how much brighter or fainter it is than each of the comparison stars shown each time you see it. Try to estimate whether it is a lot brighter or just a little brighter, and with practice you might be able to measure its brightness to within 0.1 magnitude. Its range of brightness is actually 3.5 to 4.4.

Altair and the two stars on either side of it are sometimes called the Family of Aquila. These three stars make a handy signpost to help you find other constellations, such as **Sagitta**, **Scutum** and **Capricornus**.

Brain Box

Altair spins on its axis in only 9 hours – even faster than Saturn – and measurements show that the star bulges at its equator as a result. By comparison, the sedate Sun takes over 25 days to rotate.

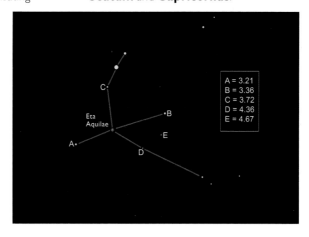

A = 3.21
B = 3.36
C = 3.72
D = 4.36
E = 4.67

Sagitta, the Arrow
M71

This tiny constellation is quite easy to find and is a great favorite among observers.

◀ M71 looks quite different from many other globular clusters, and is hardly even globular! (See the finder chart on page 169.)

Where to look

During summer and fall, when **Aquila** is in the sky, look a short way above Altair for the arrow shape of Sagitta.

Legends

There are several myths about Sagitta, but one involved a troublesome minor god, Prometheus. To punish him, Zeus commanded an eagle, Aquila, to torture Prometheus by nibbling his liver every day. His rescuer was Heracles (Hercules), who killed the eagle with this very arrow.

Point Score

●●●●●●●●●●	Date seen	Points
Finding Sagitta		2
Finding M71		3
Score		

What you'll see

The arrow consists of five stars, all of which are around fourth magnitude, making it quite tricky to see from city skies. Look with binoculars or a telescope between the two brightest stars of the shaft of the arrow and you will see a small star cluster called M71. Binoculars will just show it as a small fuzzy patch in dark skies, but with a telescope larger than about 100 mm aperture you should see some individual stars, even in less than perfect skies.

Fact File

Name	Sagitta, the Arrow			
Area	80 square degrees			

Objects	Magnitude	Distance	Type	Visibility
Gamma Sagittae	3.5	260 light years	–	–
Delta Sagittae	3.8	600 light years	–	–
M71	2.7	13,000 light years	Globular cluster	D

Vulpecula, the Fox
The Coathanger

Don't be foxed by its small size and lack of bright stars – there's plenty to see here.

Where to look

Cygnus, the Northern Cross, is a good starting point to find Vulpecula in summer and autumn skies. The bottom end of the Cross, the star Albireo, is on the northern border of Vulpecula, while the small but easily spotted **Sagitta** is at its southern edge. Between them lies Vulpecula, but because its stars are even fainter than those of Sagitta, it hardly stands out. It's a constellation with only an Alpha – no Beta, Gamma or all the rest of them. Alpha is a short distance south of Albireo.

There isn't much of a pattern other than three stars forming a flat triangle, which you'll need binoculars to find from light-polluted skies.

Legends

This is not an ancient constellation, but was one of several invented in the 17th century by the Polish astronomer Hevelius. So there are no legends linked to it. Hevelius originally called it Vulpecula et Anser, meaning "Fox and Goose." But the goose isn't there now, so maybe it has been eaten by the fox.

What you'll see

Because the stars of Vulpecula are so hard to see, the neighboring constellation of Sagitta is the easiest way to find objects in Vulpecula. One of the most attractive is the group known as the Coathanger, which is best seen using binoculars. To find it, start with the two stars marking the feather end of Sagitta and move a short distance to their northwest (up and right). You will

Point Score		
●●●●●●●●●●	Date seen	Points
Finding Alpha Vulpeculae		2
Finding the Coathanger		2
Score		

See also: **M27**.

see a line of fifth- and sixth-magnitude stars with a hook of four stars hanging down from them, making a coathanger-shape.

You can see the Coathanger with a telescope, but usually the magnification is so high that you only see a bit of it at a time.

This isn't really a star cluster, as the stars are all at different distances from us. They just appear to have this pattern as seen from our direction. This sort of thing is known as an *asterism* – a small star pattern. It is also known as Brocchi's Cluster, even though it isn't really a cluster. Brocchi was an American observer in the 1920s, who realized that this would make an easily found group of stars whose brightnesses could be measured so they could be used to help in checking other stars.

▲ The Coathanger is one of the best asterisms in the sky.

▼ When you check the distances of the Coathanger's stars, shown in light years, you realize that it is not a cluster at all.

Fact File

Name **Vulpecula, the Fox**
Area **268 square degrees**

Objects	Magnitude	Distance	Type	Visibility
Alpha Vulpeculae	4.5	300 light years	–	–
Coathanger, Brocchi's Cluster	–	–	Asterism	A
4 Vulpeculae	5.2	270 light years	–	–
5 Vulpeculae	5.6	240 light years	–	–
8 Vulpeculae	5.8	510 light years	–	–

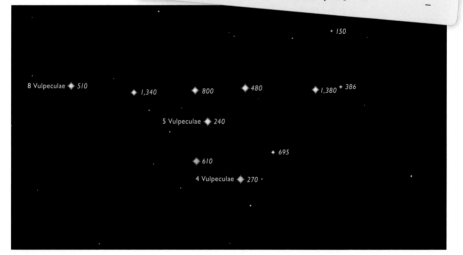

8 Vulpeculae ◆ 510   ◆ 1,340   ◆ 800   ◆ 480   ◆ 1,380   • 386   • 150

5 Vulpeculae ◆ 240

◆ 610   • 695

4 Vulpeculae ◆ 270 •

A favorite target for photographers, and the easiest planetary nebula to see with binoculars.

▶ The Dumbbell Nebula photographed using a large telescope. The two lobes that give it its name could also be an hourglass.

Fact File

Name	Dumbbell Nebula, M27		
Type		Visibility	Distance
Planetary nebula		B	1,300 light years

Where to look

Although it's in the summer and autumn constellation of Vulpecula, the easiest way to find the Dumbbell is to start from neighboring **Sagitta**. This is because the arrow pattern of Sagitta is much easier to find than the rather spread-out stars of Vulpecula itself.

Take the length of Sagitta's arrow from the feather end to the brightest star of the shaft, and imagine a right angle from the brightest star (Gamma Sagittae) up toward Cygnus. An arrow-length from Gamma lies 14 Vulpeculae. Two fainter stars make a triangle with this star, and M27 lies just below the left-hand one. The whole area is just within the field of view of typical binoculars (5°) so you don't need to look far. (Use the finder chart on page 169.)

What you'll see

With binoculars you should see a small, faint disk, which is the Dumbbell Nebula. It does need reasonably dark skies to see it, but as long as it's clear you can spot it with binoculars. With even a small telescope you should be able to see it well with a magnification of 20 or more, though 50 or 100 is better.

Although it does look like a gray disk of light to start with you might notice that it has two brighter areas within it, which gave rise to the name from their fancied resemblance to a weightlifter's dumbbell.

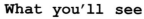

Point Score			
●●●●●●●●●		Date seen	Points
Finding M27			3
Score			

Sagittarius, the Archer
The Lagoon Nebula, M8

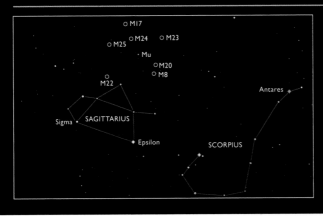

Lying at the heart of the Milky Way, Sagittarius has more deep-sky objects than any other constellation.

Where to look

Although it is quite far south, Sagittarius is well worth the effort of finding it. It is first visible in the late evening sky in May, then gets up earlier until its last gasp is at the beginning of November when it sets in the southwest. Despite being centrally placed within the summer Milky Way, Sagittarius doesn't have particularly bright stars.

From northern areas it is always low in the sky – you may need to use nearby features to help you find it. Antares is the only bright star in the area, and Sagittarius is to its left. If you can see the Milky Way, Sagittarius is to the left of the Milky Way at the same level as Scorpius.

Fortunately, the main stars of Sagittarius form a very distinctive teapot-shape.

Point Score

⬤⬤⬤⬤⬤⬤⬤⬤⬤	Date seen	Points
Finding Sagittarius		1
Seeing the Sagittarius Star Cloud		2
Finding the Lagoon Nebula		2
Seeing NGC 6530		2
Score		

See also: **M20; M22; M23–M24–M25; M17.**

Perhaps we should be grateful that such things weren't known in ancient times, or people these days might have to say that they were typical Teapots instead of typical Sagittarians, whatever that may mean! Because Sagittarius is one of the constellations of the Zodiac, it may contain a bright planet, which will outshine any of its stars.

As with several constellations, the brightest star in Sagittarius is not Alpha as you'd expect but Epsilon, followed by Sigma.

Legends

The strange thing about Sagittarius is that he is always referred to as "the Archer," which is what the word means in Latin, but in all the depictions he's shown as a centaur with a bow and arrow. A centaur is a mythical creature with the upper half of a man but the lower half of a horse. Yet there is a perfectly good centaur as an ancient constellation, Centaurus, which is best seen from the southern hemisphere. In fact, this constellation goes back even before the Ancient Greeks, to the Bronze Age Sumerian civilization of Mesopotamia (now Iraq). So there are no particular legends about Sagittarius – he's just any old archer.

What you'll see

Once you've recognized the teapot-shape you can start to pick out the various deep-sky objects – star clusters and nebulae – in the constellation. Right next to the Teapot's spout is the brightest part of the whole Milky Way, which is often called the Sagittarius Star Cloud. It is usually too low to be seen well from northern areas.

This is a great area to just scan with binoculars on a good night, as there are

▲ The Lagoon Nebula is one of the showpieces of the sky, though visually it is a bit disappointing. The cluster of stars, called NGC 6530, to the east of the brightest patch of the nebula is the easiest part to see in poor skies.

countless stars, lots of little clusters, knots of stars and misty patches. You are looking right toward the center of the Milky Way, but oddly enough the actual center of the galaxy is hidden by a fairly nearby dark cloud and we can't see it directly.

Above the star marking the spout of the Teapot is a very obvious but smaller misty patch, which is the Lagoon Nebula, M8. This is easily visible to the naked eye on a good night. It is a bright nebula, like the Orion Nebula. In photos it appears red, but our eyes are not sensitive enough to see this and it appears pale and translucent.

Fact File

Name Sagittarius, the Archer
Area 867 square degrees

Objects		Magnitude	Distance	Type	Visibility
Kaus Australis	Epsilon Sagittarii	1.8	143 light years	–	–
Nunki	Sigma Sagittarii	2.1	227 light years	–	–
M8		–	4,300 light years	Gaseous nebula	A

173

The Trifid Nebula, M20

Its picture is in almost every astronomy book and it has a great name, but the **Trifid Nebula** can take some finding.

▶ The Trifid Nebula looks much better in photos than it does in a telescope!

Where to look

Find Sagittarius, and from there look for **M8**, the Lagoon Nebula, above the star marking the end of the Teapot's spout. A short distance above M8, and well within the same field of view of binoculars, is the Trifid Nebula, M20. (Use the finder chart on page 172.)

What you'll see

Probably not much! In average skies it looks like a hazy star, with the haze being rather smaller than that of the Lagoon Nebula. To see it well you'll need a telescope and good skies. Pictures show a beautiful double nebula, with one part red and

Fact File

Name	Trifid Nebula, M20		
Type		Visibility	Distance
Gaseous nebula		E	2,700 light years

the other part blue. The red part has three dark spokes within it, which gave the nebula its name Trifid – meaning "divided into three." There is also a famous book and film about man-eating plants called Triffids, but that's completely different! The blue color is starlight reflected from dust, while the red is the color that hydrogen gas glows in the light from a nearby hot star.

Because it's much fainter than the Lagoon Nebula it is much harder to see, but the fun comes from finding this object which is so often seen in photos. To see the three dark spokes you'd need a good telescope and crystal clear, dark conditions. But you won't see the colors, which only show up in photographs.

Point Score

● ● ● ● ● ● ● ● ● ●	Date seen	Points
Seeing the Trifid Nebula		5
Score		

M22

One of the brightest globular clusters in the sky, yet it's often overlooked because it's so far south.

◄ If you have really good skies, or live far enough south, globular cluster M22 is a rival for M13. This photo was taken from Kitt Peak Observatory in Arizona.

Where to look

Find the Teapot of Sagittarius, and look just above and to the left of its top star to find this object. With binoculars it should be well within the same field of view. (Use the finder chart on page 172.)

What you'll see

Even small binoculars show a fuzzy blob here, assuming that the area isn't too low or hidden by light pollution. With a small telescope you can start to pick out individual stars, and with even a 130 mm telescope it starts to look quite spectacular. It is the third brightest globular cluster in the whole sky – the others being

Point Score

Point Score	Date seen	Points
●●●●●●●●●● Finding M22		3
Score		

Omega Centauri (visible from southern states) and 47 Tucanae (visible from the southern hemisphere only).

So why isn't it better known, and ranked higher than the "Great Hercules Cluster," M13, in lists of things to view? The answer is that being so low in the sky, its light is dimmed so much that it's less spectacular. Hercules can appear virtually overhead, which makes a big difference. But M22 is a lot closer – less than half the distance of M13, and one of the closest of all globulars.

The trick is to choose your moment, and wait for the very best conditions. If possible, go to the darkest site you can, and you should be rewarded with a splendid view.

Fact File

Name M22

Type	Visibility	Distance
Globular cluster	A–C depending on location	10,400 light years

175

M23–M24–M25

This chain of star clusters in Sagittarius is an easy target on a good night, if you have a good southern horizon.

Fact File

Name	Type	Visibility	Distance
M23	Open cluster	C	2,000 light years
M24	Star cloud	D	Various
M25	Open cluster	C	2,000 light years

▲ M25, M24 and M23 run in an east–west line, though M24 is not really a cluster like the others but a brighter part of the Milky Way.

Where to look

First look some way above the lid of the Teapot of Sagittarius to find a brighter part of the Milky Way, the Little Sagittarius Star Cloud, aka M24. On either side of it are M23 and M25, each one about 4° away. (Use the finder chart on page 172.)

What you'll see

Cluster M25 consists of a handful of stars scattered over an area of sky about the size of the Moon. It's the easiest of the three objects to see from a light-polluted area.

Now move west to M24, which is much larger than either of the two others,

covering an area of 2° × 1°. This is not really a cluster but a star cloud that just happens to be bright, so there is no particular distance for it.

Another shift west and you are on to M23, which has more but fainter stars than M25 so it's a challenge in poor skies. It reminds some people of a spider – what about you?

Point Score

○○○○○○○○○○	Date seen	Points
Finding M23		3
Finding M24		4
Finding M25		3
Score		

The Swan Nebula, M17

It's got several names – so take a look for yourself and decide what to call this nebula.

▶ Through even a small telescope you can see the swan-shape of M17 very clearly, though as always there are no colors visible to the eye. This digital camera photo was through an 80 mm refractor with an exposure time of 2 minutes and has north at the top so the swan is upside down.

Where to look

If you can find **M24**, the Little Star Cloud in Sagittarius, get the left-hand side of it in your binoculars and M17 will be in the same field of view, above it. But if M24 is too faint for your skies, don't give up, because M17 is brighter. Instead, find the lid of the Teapot and use it as an arrow to point to a faintish star, Mu Sagittarii, about a binocular field above the lid. This is just below M24. Now scan upward and to the left, about another binocular field of view, and M17 should appear. It is actually due north of the top lid star. (Use the finder chart on page 172.)

What you'll see

M17 is a gaseous nebula, like M8 and M20, and like those it appears as a little patch of mist or cloud, pale and colorless even though many photos show it as red. Each of these nebulae has its own distinctive shape, and M17 is a triangle or wedge. But

from one side there is a hook-shape of gas, which has resulted in different names from different people. Some Victorian astronomers, who learned Greek at school, called it the Omega Nebula because it looks like a Greek capital Omega. Others, maybe more down-to-earth, saw it as a horseshoe. It also looks like the number 2, and is also called the Checkmark Nebula and the Lobster Nebula. But it also makes a pretty good swan, so you can make up your own mind.

Fact File

Name	Swan Nebula, M17	
Type	Visibility	Distance
Gaseous nebula	C	5,900 light years

Scutum, the Shield
M11

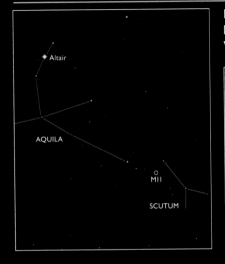

It's one of the smallest constellations but it contains a deep-sky gem: the Wild Duck Cluster.

▲ The Wild Duck Cluster is named because of a V-shape of its brighter stars, which show up fairly well in this photo.

Where to look

Begin by finding **Aquila**, which has a curve of stars starting with the line of three of which Altair is a part. Follow this curve southward and you come to a smaller C-shaped group of stars. The top of this little pattern is in Aquila, but the bottom of it is in Scutum.

Legends

Scutum is not an ancient constellation, but was invented in the 17th century. In this case it was the Polish astronomer Hevelius who devised it to honor his patron, King John

Fact File

Name	Wild Duck Cluster, M11		
Type		Visibility	Distance
Open cluster		C	6,100 light years

Point Score

●●●●●●●●●●	Date seen	Points
Finding Scutum		2
Finding M11		3
Score		

Sobieski. It was originally Sobieski's Shield, but now we drop the sponsorship bit.

What you'll see

Look just below the bottom of the C-shape with binoculars or a telescope on a good night and you'll spot a quite small star cluster, known as M11, the Wild Duck Cluster. Its stars are mostly 10th magnitude, so you'll need fairly dark conditions to see it at all, but it is rich in faint stars and is best seen in a telescope with a magnification of at least 50.

OBJECTS TO LOOK FOR IN AUTUMN

With the autumn (also known as the fall) come darker nights that are warmer than those of spring. And many of the summer objects are still around for another look.

The nights are drawing in so the all-sky map on this page applies for November 1 at 8.00 pm, but you can recognize the same constellations a month or two on either side of that or earlier or later in the evening.

The bright star Vega is high up toward the west, and the large Summer Triangle of Vega, Deneb and Altair is still a good starting point. But now we have the Square of Pegasus high in the south, so the sky is full of geometrical shapes to help you.

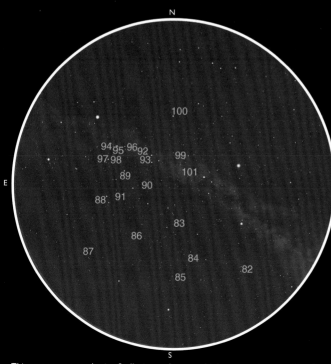

This map was created using Stellarium software, which you can download free from www.stellarium.org.

Capricornus, the Water Goat
Alpha Capricorni and Beta Capricorni

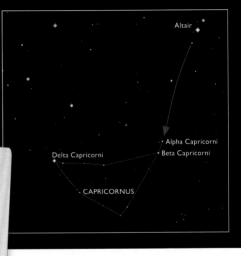

Test your eyes and your telescope on the double stars in this part of the sky.

Fact File

Name Capricornus, the Water Goat
Area 414 square degrees

Objects	Magnitude	Distance
Alpha1 Capricorni	4.3	568 light years
Alpha2 Capricorni	3.6	106 light years
Beta1 Capricorni	3.1	328 light years
Beta2 Capricorni	6.1	328 light years

Where to look

Capricornus is fairly close to the horizon in late summer and fall. The easiest way to find it is to locate the bright star Altair, which is the one with two stars on either side of it, and is part of the Summer Triangle. Follow the line of the three stars down toward the horizon, veer right a bit, and you come to Alpha Capricorni, with Beta a short way below it on the same line. Beta is the brighter of the two.

Whether you can pick out many of the other stars of the constellation depends on how good your skies are. The only other bright star is Delta, some way to the east, which is about the same brightness as Beta, and the other stars form a broad triangular shape pointing downward.

Because the constellation is on the ecliptic, there could be a bright planet there as well.

Legends

You might think that only Greek legends could invent a water goat. But this odd creature actually dates back earlier than the Ancient Greeks, to the Bronze Age civilizations of Mesopotamia (now Iraq).

It wallows in the watery area also inhabited by Aquarius, Cetus and several fish (see the article on **Pisces** for the reason why this is such a soggy part of the sky).

In fact, the Greeks took over the fish-tailed goat and instead regarded him as Pan, who was half man and half goat. Pan was a nature god who invented a musical instrument made from reeds of different lengths, which we now call the pan pipes. There are dozens of tales about the exploits of Pan, but in one of them he dived into the Nile to escape a monster, turning his lower half into a fish. In this guise he helped to save Zeus, the king of the gods, from the monster, thus earning his place among the stars.

What you'll see

The stars of Capricornus are mostly rather faint, but turn your attention to Alpha Capricorni. Most people can see that this consists of two stars, of roughly equal brightness, separated by over 6 minutes of arc. But despite their similarity, these two stars are completely unconnected. The slightly fainter one is about five times farther away than the brighter one.

▲ Alpha Capricorni (top) and Beta Capricorni. The field of view of this photo is about the same as 10 × 50 binoculars.

However, look through a telescope and you'll see that each star is accompanied by another, fainter star, making it a double-double star. But these stars are probably not connected with their apparent partners.

Also look for

Beta Capricorni is also a double star, though you need binoculars to see the two stars separately. The two stars are genuine partners rather than being one in front of the other at different distances, but they are so far apart from each other that they will take maybe 700,000 years to orbit each other.

▲ Alpha Capricorni in close-up as seen through a small telescope. A low-power eyepiece is all you need for this view.

Point Score		
●●●●●●●●●●	Date seen	Points
Finding Alpha and Beta Capricorni		1
Finding other stars in Capricornus		1
Seeing Alpha as a double star		1
Seeing Beta as a double star		2
Score		

Pegasus, the Winged Horse
M15

The stars of Pegasus are a very useful guide to finding your way round autumn skies.

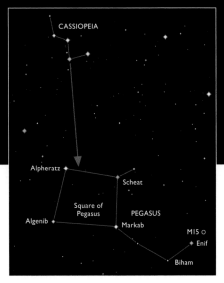

Where to look

Pegasus is quite high up in the sky looking south from late at night in September until early evening in January, and is either rising or setting either side of those months. Whenever the W-shape of **Cassiopeia** is high in the sky, you can find Pegasus. Follow the line of the third and fourth stars of the "W" (counting from left to right) and you'll find it. This is actually a rather barren patch of sky, so although the stars are not especially bright, they are easy to pick out. Look for a large rectangle, known as the Square of Pegasus.

Point Score		
●●●●●●●●●	Date seen	Points
Finding Pegasus		1
Counting stars in the Square of Pegasus		2
Finding M15		3
Score		

Legends

Pegasus was a mythical winged horse, born from the body of the Gorgon Medusa when Perseus slew her. In time he became the steed of the hero Bellerophon, and together they went into battle against yet another dreadful monster, the Chimera, a creature with the head of a lion, the body of a goat and the tail of a serpent. Bellerophon felt that killing the Chimera made him worthy of a place on Mount Olympus, where the gods lived, and flew there aboard Pegasus. But this was too much for Zeus, who allowed Pegasus a place in the stars while you won't find Bellerophon anywhere these days.

Fact File

Objects		Magnitude	Distance	Type	Visibility
Name	Pegasus, the Winged Horse				
Area	1,121 square degrees				
Markab	Alpha Pegasi	2.5	133 light years	–	–
Scheat	Beta Pegasi	2.4	195 light years	–	–
Algenib	Gamma Pegasi	2.8	400 light years	–	–
Enif	Epsilon Pegasi	2.4	700 light years	–	–
M15		–	33,000 light years	Globular cluster	C

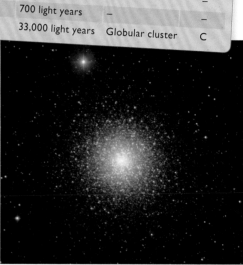

▶ This photo of M15 with a large telescope shows how densely packed are stars in a globular cluster.

In the sky, Pegasus is shown with only his front half, though luckily this includes the wings. What's more, he is upside down, so the bit that represents the mane and head is below and to the west of the Square of Pegasus that marks his body.

What you'll see

The Square of Pegasus is not actually a perfect square, but let's face it, we have to put up with the stars as they are and not how we would like them to be. The "Quadrilateral of Pegasus" just wouldn't catch on. All four stars are roughly the same brightness and are visible from all but the worst skies. The Square might be a bit larger than you were expecting – it is about the width of your outstretched hand at arm's length. The top left star, Alpheratz, is part of the Square but it isn't actually in Pegasus but in neighboring Andromeda.

The neck of the horse extends from the lower right star in a line down to a star called Biham, and then bends upward again to the star Enif, marking his nose. His front legs stick upward from the top right star.

The Square of Pegasus can be used for checking your eyesight, or for working out how good your skies are. Just count the number of stars within the Square, not including the four that make up the Square. There's just one star brighter than magnitude 4.5, four brighter than 5.0, seven brighter than 5.5, and 13 brighter than magnitude 6.0. So how faint a star can you see?

Looking down at Pegasus's neck, if you continue the line from Biham to Enif half as far again you come to the globular cluster M15, a little fuzzy ball just to the right of a sixth-magnitude star. With small binoculars and in poor skies you will be hard pressed to see it, but it gets better as the skies get darker and you use more magnification. With a telescope of 100 mm or larger you can begin to see stars in it, but it is more distant than most of the other bright globular clusters.

183

Aquarius, the Water Carrier M2

Everyone's heard of Aquarius, but it isn't easy to spot as its stars are mostly quite faint.

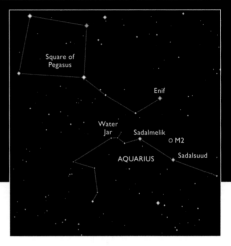

Where to look

As with most constellations in this part of the autumn and winter sky, the best place to start is the Square of Pegasus, which is high up at this time of year. From there, find the star Enif which marks the nose of the winged horse, and then look below it for a star of about the same brightness, as far below it as the width of the Square. This is Sadalsuud, very slightly the brightest star in Aquarius.

From there, move toward the Square to find a slightly fainter star, Sadalmelik, which is Alpha Aquarii. To the left of that lies a little pattern of stars which is the most recognizable feature of the whole constellation, a sort of arrow-shape of four fourth-magnitude stars. In poor skies you'll need binoculars to see it. This pattern is known as the Water Jar, but most people these days would see it as an aircraft with wings swept back.

The other stars of Aquarius are scattered around the area, not forming

Fact File

Name	Aquarius, the Water Carrier				
Area	980 square degrees				
Objects		Magnitude	Distance	Type	Visibility
Sadalmelik	Alpha Aquarii	3.0	435 light years	–	–
Sadalsuud	Beta Aquarii	2.9	535 light years	–	–
M2		–	18,000 light years	Globular cluster	B

any obvious shape and, as with Capricornus, you need good skies to be able to see them at all.

Legends

Probably the reason why Aquarius is called the Water Carrier dates back long before the Ancient Greeks told tales about the patterns in the sky. For the explanation why this area is linked with water pouring from the sky, see the article on **Pisces**.

There is no classical figure known as Aquarius, but the Greeks linked it with the beautiful boy Ganymede, who was a favorite with several of the gods and who was plucked from the Earth by the eagle Aquila. While the stories about Ganymede have nothing to do with water, he did become cupbearer to the gods – a sort of celestial wine waiter.

What you'll see

The brightest deep-sky object in Aquarius is the globular cluster M2. This sits by itself with no bright stars near it, but fortunately it is almost exactly due north from Sadalsuud. So find this, then move 5° north (about one binocular field of view) and you should

Point Score

●●●●●●●●●●	Date seen	Points
Finding Sadalsuud, Sadalmelik and the Water Jar		2
Finding other stars in Aquarius		2
Finding M2		2
Score		

spot it as a rather fuzzy star. It's one of the brighter globular clusters, but it's quite distant so you need at least a 100 mm telescope and a magnification of more than 100 to see individual stars in it.

▲ The Water Jar is the most recognizable part of Aquarius, and helps you to find the rest of the constellation.

◀ The globular cluster M2 is one of the easier globular clusters to find as it is north of Sadalsuud.

185

The Saturn Nebula and the Helix Nebula

These two famous planetary nebulae are a challenge for binoculars and small telescopes.

Where to look

First, the Saturn Nebula, for which you'll need a telescope. Begin by finding Sadalsuud (see page 184, **Aquarius**). From there, you need to find a fainter star, Nu Aquarii, which is fourth magnitude. This is about 8° to the southwest of Sadalsuud, and there is a flat triangle of stars to its west which helps you to find it by extending one of the sides of the triangle. Having found Nu Aquarii, use a magnification of at least 50, and preferably

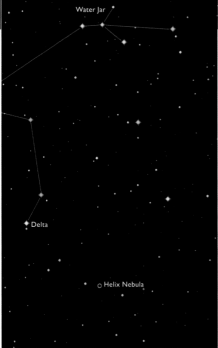

Point Score									Date seen	Points
Finding Saturn Nebula										4
Finding Helix Nebula										10
Score										

more, and look for a star about a degree to its west. The thing to look for is its color. While most stars are white, pale red or possibly pale blue, the Saturn Nebula is noticeably blue-green.

The Helix Nebula is a different animal altogether. It requires the very darkest of skies, particularly as it is very low down in the sky as seen from northern areas,

and usually binoculars rather than a telescope. It's in a blank area of sky, so there are no nearby bright stars to help you find it. The nearest signpost is the third-magnitude star Delta Aquarii, which has a fainter star just below it. From there, you just have to use the finder chart to get yourself to the right spot in the sky, midway between two fifth-magnitude stars.

▼ It is possible to see the Helix Nebula with binoculars – I have seen it with 10 × 30s – but only in the darkest skies. But this is what it looks like with the Hubble Space Telescope.

Fact File

Name	Visibility	Distance
Saturn Nebula, NGC 7009	C	2,400 light years
Helix Nebula, NGC 7293	E	715 light years

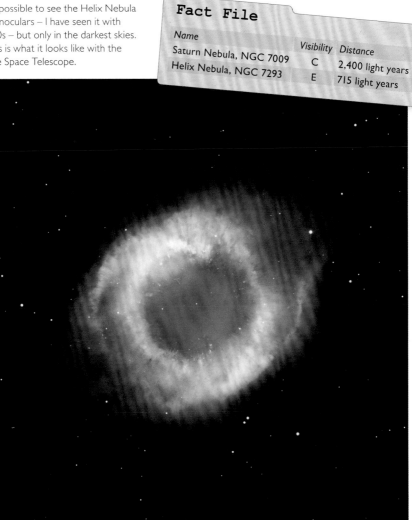

What you'll see

In the case of the Saturn Nebula, the striking thing is its color. It is actually about the same size as the planet Saturn, which is why low magnifications just show it as a star. With a magnification of 75 or 100, you can see a blue-green, slightly oval disk. But the resemblance to Saturn only comes with very large telescopes, or in photos, which show faint wings on either side of the nebula that look a bit like the rings of Saturn.

Even if you have found the right place for the Helix Nebula, the chances are that you will see nothing unless you have pitch-black skies in a remote country area. The slightest bit of light pollution from a town miles away is enough to drown it out. You need to look for a circular hazy patch, about a third of the diameter of the Moon. If you have an O III filter for your eyepiece,

Brain Box

The Hubble photo of the Helix Nebula has been called the "Eye of God" because of its shape, but some internet hoaxers claim that it can only be seen every 3,000 years! You can prove them wrong by finding it yourself.

that may make it a bit easier to see, but the filter does dim the other stars and the sky so at first glance all you see is blackness.

But the funny thing is, this is just about the closest planetary nebula to Earth! So how come it's so hard to see? Its light really is spread out over a wide area of sky, so it is very little brighter than the background. For most people this is something to aim for on that special occasion when you find yourself far from city lights. But photos of it are fantastic, so it is worth looking for just to say you've tried.

Finding the distance to planetary nebulae is tricky. The best and most recent estimate of the distance to the Helix comes from measuring its position as seen from opposite sides of Earth's orbit (called *parallax*) and gives 715 light years – but only to limited accuracy. So it could be between 645 and 800 light years away.

◀ The Saturn Nebula is worth looking for because of its strong color, but don't expect to see the faint extensions which make it look like the planet.

86

Pisces, the Fishes
The Circlet

Like all fish, these two can be very elusive and are hard to catch, unless you have good skies.

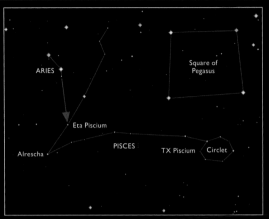

Where to look

Pisces is an autumn constellation, and the best way to find it is to start with the Square of Pegasus high up on autumn evenings. To the left of the Square, about 1½ Square-widths away, are the main stars of Aries. These are two brightish stars with another fainter one at an angle. The brighter pair point southwestward to the brightest star in Pisces, called Eta Piscium.

From here you might be able to trace a line of faint stars leading southeast to northwest, which marks one of the fish.

At the bottom of the line is Alrescha, Alpha Piscium, and from there you can look for another line of faint stars ending up below the Square of Pegasus.

Do you say fish or fishes? Most people translate the name as "Fishes," but in everyday speech we normally use just fish for the plural.

Legends

Fish are not generally reckoned to be great heroes, but these two are said to have rescued the goddess Aphrodite and her son Eros (also known as Venus and Cupid) who leapt into a river to escape a dreadful monster. Oddly enough, the two fish are tied by their tails, and Alrescha marks the knot in the cord that binds them.

This part of the sky is very watery. The river Eridanus lies nearby, with the sea monster Cetus below the fish and

Point Score

●●●●●●●●●●	Date seen	Points
Finding Alpha and Eta Piscium		2
Finding the Circlet		2
Finding more stars in Pisces		2
Score		

the water goat Capricornus to their west, and the Southern Fish below that. The water carrier Aquarius lies beyond. In the parched lands of Mesopotamia, where many constellations originated, October or November marked the start of the rainy season, and still do today. The stars that are in the evening sky at that time of year all have watery connections.

What you'll see

Trace out the stars of Pisces and you'll find that right below the Square of Pegasus is a ring of stars. This is known as the Circlet,

and it marks the head of one of the two fish. It's faint, so you'll need binoculars unless you have a dark sky, but it's one of the most distinctive features of the constellation. It's an asterism, not a cluster, as the stars are all at different distances from us.

The easternmost star of the Circlet is a particularly red star called TX Piscium. Take a look at it with binoculars or a telescope, as it's about as red as stars can get, but don't expect it to look as red as a brake light. It is a variable star – that is, one whose brightness varies.

▼ In taking this photo of the Circlet (in the lower half of the picture), I put a diffusing filter over the lens to smear out the star images, which helps to show the star colors.

Fact File

Name	Pisces, the Fishes		
Area	889 square degrees		
Objects		Magnitude	Distance
Alrescha	Alpha Piscium	3.8	150 light years
	Eta Piscium	3.6	350 light years
	TX Piscium	4.9–5.5	850 light years

87

Cetus, the Sea Monster
Mira, the Wonder Star

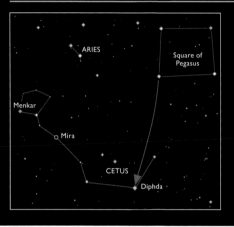

The star Mira amazed astronomers when they first spotted it – see if you can find it, too.

▲ With the Hubble Space Telescope, the disk of Mira appears just a few pixels across. It changes shape from time to time, the hook of gas probably being drawn off by a companion star.

Where to look

Cetus is around when Pegasus is high in the sky, and the best time to see it is in evenings from November to January.
It lies some way closer to the horizon than Pisces, and has few bright stars so you may have trouble finding it. Its brightest star, Diphda, is below the Square of Pegasus, and marks the tail of the creature. Follow the line of the left-hand side of the Square downward by a couple of lengths, and cheat a bit by going a bit farther to the left, and you come to this second-magnitude star sitting more or less by itself in a fairly blank bit of sky. The ecliptic runs through this area, so any bright star nearby is going to be a planet.

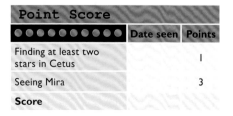

Point Score		
●●●●●●●●●	Date seen	Points
Finding at least two stars in Cetus		1
Seeing Mira		3
Score		

The other end of this large constellation is below Aries. Draw a line at right-angles to the two brightest stars of Aries and you come to a third-magnitude star, Menkar, at one end of a triangle of fainter stars. There's a ring of fainter stars here marking the head of the sea monster. Between here and Diphda is a straggle of faint stars. Or at least, usually there are only faint stars, but read on to find out why this isn't always true.

Legends

This was the sea monster that in the legend of Andromeda and Perseus was about to eat Andromeda when Perseus rescued her. Traditionally this was a proper ugly monster, quite unlike any normal sea creature, but these days the word "cetacean" applies to whales, so some lists call Cetus a whale. But, as we know, whales don't usually eat people.

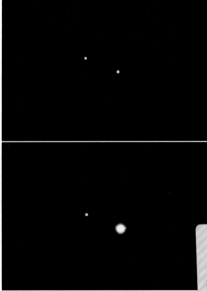

▲ Close-up of Mira when faint (top) and bright (below). The left-hand star is not connected with Mira and is magnitude 9.5.

What you'll see

Every so often, there is an extra star in the body of Cetus. This is the star Mira, and occasionally it is brighter than Diphda, but for most of the time it is invisible. To find it, go to the triangle of stars in the head of Cetus and draw a line from Menkar through the opposite corner of the triangle. If Mira is bright, it lies about the same distance again along the line. But the chances are that you won't see anything there.

The reason is that Mira is a variable star, whose brightness rises and falls on a fairly regular basis. It takes about 330 days to go from one bright stage to the next, and remains fairly bright for only about three months of the year unless you really search for it. Then it drops out of sight again, and you need binoculars to find it at all.

Because the interval from peak to peak, or period, of the star is 11 months, if it happens to be bright when it is in the daytime sky you won't see it for several years. So in the years up to 2017 it will be tricky to see Mira bright, but from then until about 2021 you stand a good chance as it is brightest in winter when it's in the evening sky.

Mira was the first star that was noticed to vary in brightness over a period of time, and in the 17th century this was considered so remarkable that it was given the name Mira, meaning "wonderful" in Latin. Its variations are caused by pulsations in its size. Mira is a huge star, and is one of few stars that show a disk when seen with very large telescopes.

Fact File

Name	Cetus, the Sea Monster		
Area	1,231 square degrees		
Objects		Magnitude	Distance
Menkar	Alpha Ceti	2.5	250 light years
Diphda	Beta Ceti	2.0	100 light years
Mira	Omicron Ceti	2.5–10	300 light years

Aries, the Ram
Mesartim

Famous for being a sign of the Zodiac, Aries is easy to spot and contains an easy double star.

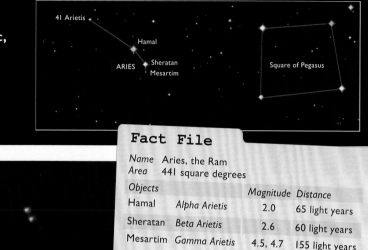

41 Arietis

Hamal

ARIES Sheratan
Mesartim

Square of Pegasus

Fact File

Name	Aries, the Ram		
Area	441 square degrees		

Objects		Magnitude	Distance
Hamal	Alpha Arietis	2.0	65 light years
Sheratan	Beta Arietis	2.6	60 light years
Mesartim	Gamma Arietis	4.5, 4.7	155 light years

▲ Mesartim is a double star with two well-separated, blue-white stars that can easily be seen with a small telescope, though not binoculars.

Where to look

Aries is a constellation of late fall and winter evenings. Begin with the Square of Pegasus and look about 1½ Square-lengths to its east (left) for the main star of Aries, Hamal, with a fainter star, Sheratan, a few degrees to its right. Just below that is a third star, Mesartim, at an angle to the other two.

Point Score

⚬⚬⚬⚬⚬⚬⚬⚬⚬⚬	Date seen	Points
Finding Aries		1
Seeing Mesartim as two stars		2
Score		

Legends

This ram is famous for providing a golden fleece which became involved in other stories. The ram had wings, and rescued the two children of Nephele, the goddess of clouds, when the boy, Phrixus, was destined to be sacrificed. But his sister Helle fell from the flying ram's back, and was drowned in the strip of water between Europe and Asia. This became known as the Hellespont in her honour. In the end the poor old ram was himself sacrificed by Phrixus, who should have been more grateful, but at least the ram has been placed in the stars.

What you'll see

Aries has no bright deep-sky objects, but Mesartim is a double star whose two stars are separated by just over 7 seconds of arc and take about 5,000 years for one orbit.

Andromeda
NGC 752

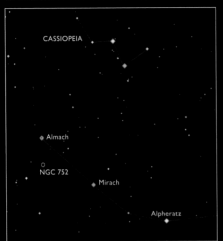

CASSIOPEIA

Almach

○
NGC 752

Mirach

Alpheratz

Andromeda is famous for its galaxy, but there's a whole constellation to see as well.

Where to look

Andromeda is best seen on late fall and winter evenings. First find the "W" of **Cassiopeia**, more or less overhead, then look in the direction that the points of the "W" indicate. You'll see a line of three widely spaced bright stars, the westernmost of which is part of the Square of Pegasus. These are the main stars of Andromeda, but there is a considerable part of the constellation with no bright stars to the southwest of Cassiopeia as well.

Legends

Andromeda was the daughter of the extremely vain queen Cassiopeia, who was always bragging about something or other. Eventually she went too far, by claiming that Andromeda was more beautiful than the Nereids, who were the female spirits

Point Score

●●●●●●●●●	Date seen	Points
Finding Andromeda		1
Seeing Almach as a double star		2
Finding NGC 752		2
Score		

See also: **Andromeda Galaxy.**

of the sea. As a punishment, the blameless Andromeda was chained to a rock on the coast. The sea monster Cetus would have eaten her if it hadn't been for the timely arrival of Perseus, fresh from slaying Medusa, who rescued her.

What you'll see

As well as the line of the three main stars, Alpha (Alpheratz), Beta (Mirach) and Gamma (Almach), look for a fainter star

▲ Double star Gamma Andromedae (Almach) is one of the most striking in the sky and is easily viewed with a small telescope.

Your binoculars will come in handy for the next object, NGC 752, which is just 5° south of Almach, so you can put Almach at the top of the field of many binoculars and NGC 752 will be near the bottom. It is a star cluster, visible in binoculars. You'll see a scattering of stars, mostly magnitude 8 or 9, covering an area a bit larger than the full Moon. Because this cluster is quite large, you might get a better view with binoculars than with a telescope.

roughly between Alpha and Beta, and another line of fainter stars that lies to the north of these, branching off from Alpha.

The star Gamma Andromedae, Almach, is well worth a look through a telescope. It's a double star, like Albireo in Cygnus and Mesartim. In this case the main star is a lovely orange color, while the fainter secondary star is bluish. The two are separated by 10 seconds of arc, which is about a quarter of the diameter of Jupiter, so most binoculars don't have enough magnification to see the two separately. But with a telescope a magnification of about 50 should be enough.

▲ Often overlooked, but easy to see with binoculars as well as telescopes, NGC 752 is an attractive star cluster in Andromeda.

Fact File

Name Andromeda (personal name)
Area 722 square degrees

Objects		Magnitude	Distance	Type	Visibility
Alpheratz	Alpha Andromedae	2.1	100 light years	–	–
Mirach	Beta Andromedae	2.1	200 light years	–	–
Almach	Gamma Andromedae	2.1, 4.8	350 light years	–	–
NGC 752		–	1,500 light years	Open cluster	C

90

The Andromeda Galaxy, M31

Our sister galaxy gives us the chance to see what the Milky Way would look like from outside.

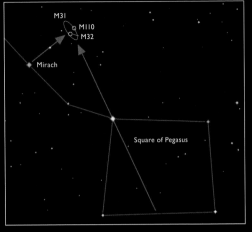

Where to look

Starting from the Square of Pegasus, count two stars along from the top left corner to get to Beta Andromedae (Mirach), then two stars upward, and you can see the Andromeda Galaxy near the uppermost star. Alternatively, draw an imaginary line from the halfway point of the bottom of the Square through the top left star and continue it an equal distance, and you've found the same spot.

The galaxy is visible from late on July evenings right through the autumn and winter until March, when it gets low in the west.

What you'll see

If you have a reasonably dark sky you can often see the Andromeda Galaxy with the naked eye, without any binoculars or telescope, by using the trick of averted vision (see page 110). It's amazing how this pale fuzzy patch comes into view if you just look slightly away from where you know it to be. You can see that it isn't circular but is elongated in the direction of the Square of Pegasus.

Binoculars or a telescope give a better view, though you'll need to use the lowest magnification of your telescope for the best results. It might even fill the whole field of

Point Score	Date seen	Points
● ● ● ● ● ● ● ● ● ●		
Finding M31		2
Viewing M31 through binoculars or a telescope		1
Viewing M32 through binoculars or a telescope		2
Viewing M110 through binoculars or a telescope		2
Score		

view, so moving the telescope slightly from side to side could help you to work out what is galaxy and what is sky.

Fact File

Name	Type	Visibility	Distance
Andromeda Galaxy, M31	Galaxy	A	2,500,000 light years
M32	Galaxy	C	2,500,000 light years
M110	Galaxy	C	2,500,000 light years

In fact, you are probably looking only at the central part of the galaxy – the nucleus. Those spiral arms are usually too faint to be seen with small telescopes in average conditions. The darker your sky, the more you will see of the extent of the galaxy.

Look around the edge of the galaxy with a telescope and you may see two smaller glows, which are companion galaxies to M31. The difference in size is because they are genuinely smaller galaxies at virtually the same distance as M31, rather than being because they are much more distant. The closer one is circular and is M32, while the more distant one is elliptical and is known as M110. You can even see these with high-power binoculars.

In photos, M32 often appears within the glow of M31 itself, but when you see it through a telescope it is some distance away. This shows you how much of M31 you are missing when observing visually.

The Andromeda Galaxy may be a sister galaxy to the Milky Way, but the two are not identical. It is larger than the Milky Way, though just how much bigger is uncertain as astronomers have to rely on estimates. The best guess is that the Milky Way contains 200–400 billion stars, while M31 contains 1,000 billion stars (a trillion).

Even so, it is quite similar, and anybody looking at the Milky Way from M31 would see it looking more or less the same in their skies as M31 does in ours. You might like to wonder, when you observe M31, about how many alien eyes are looking at you across the gap between the two. But bear in mind that light takes 2,500,000 years to travel the distance. The aliens are seeing our galaxy as it was before the human race even came into existence, when our distant ancestors were trying to avoid being a lion's lunch.

The two galaxies are the largest members of the Local Group of galaxies, which has a few dozen members. M32 and M110 are also in the group, as well as **M33 in Triangulum** and the two Magellanic Clouds, which are in the southern hemisphere and are not visible from North America. Most galaxies seem to occur in groups, with large empty spaces or voids in between.

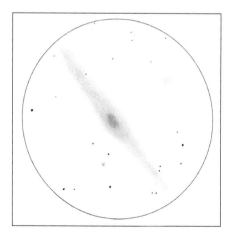

▲ A sketch of the Andromeda Galaxy using 15 × 70 binoculars from Burnley, Lancashire, UK.

▼ This pair of photos of the Andromeda Galaxy show the difference between the visual view and a long-exposure photo. Below is a 1-minute exposure through a 200 mm telescope, which is similar to the visual view, except that the center of the galaxy is brighter than you'd see. At bottom is a combination of nearly 3 hours of similar exposures. The visual view shows only the central part of the galaxy, with M32 some distance away. But in the photo, M32 is seen through the outer spiral arms.

Photo versus eye

Looking at the photos of the Andromeda Galaxy, you might be disappointed that the view through your telescope doesn't look as detailed. Would a bigger telescope give a grander view? Probably not, because photos always bring out more detail in these deep-sky objects. In a really dark sky, and with a telescope of about 400 mm aperture, you can start to see structure in spiral galaxies such as M31, but they still appear pale and transparent rather than bright and shiny.

The visibility of M31 is more or less the same as that of the Milky Way. This isn't just a coincidence – both are galaxies, and we see the Milky Way from inside and M31 from outside. Artists like to produce beautiful illustrations of galaxies seen from nearby planets, showing all the details and colors that you can see in a photo, but the sad truth is that a galaxy, whether the Milky Way or M31 or even a giant elliptical galaxy, wouldn't appear very much brighter if you were close to it than it does in the sky – just larger.

91

Triangulum, the Triangle
M33

Finding the nearby Pinwheel Galaxy, M33, is a nice challenge if your skies are dark enough.

Fact File

Name	Triangulum, the Triangle		
Area	132 square degrees		
Object	Type	Visibility	Distance
M33	Galaxy	C	2,900,000 light years

Where to look

The autumn and winter constellation of Triangulum is in the sky at the same time as **Andromeda**, so that's the best starting point. Start by finding Gamma Andromedae (Almach), then look for the distinctive pattern of **Aries**, just to its south. Triangulum lies between the two, consisting of a long triangle of fairly faint stars that should just be visible from even suburban skies.

Legends

Though Triangulum is an ancient constellation, known to the Greeks, there are no legends associated with it. It's just a triangle!

What you'll see

Apart from the three stars, the most interesting thing in Triangulum is the galaxy M33, also known as the Pinwheel Galaxy. The easiest way to find it is to use one of the same procedures as for finding M31,

Point Score

	Date seen	Points
Finding Triangulum		1
Finding M33		3
Score		

Group of galaxies, but it is smaller and more distant, so it appears fainter. It is also face-on to us, which spreads its light out over a wider area of the sky. So you need to look for a fairly large and faint glow of light, roughly the same size that the Moon appears to be. As usual, you won't see any signs of the spiral arms or details, even with a telescope, unless you have really clear and dark skies, well away from any cities.

▲ A long-exposure photo brings out the spiral arms and the pink hydrogen gas clouds within them – star-birth areas similar to the Orion Nebula in our own galaxy.

which is to go to Mirach, Beta Andromedae. To find M31 you count two stars upward, but M33 is an equal distance downward from Mirach, at right-angles to the line of stars in Andromeda. While M31 is visible with the naked eye, for M33 you will need binoculars.

Like the Andromeda Galaxy, this is a member of the Local

▲ A photo of M33 through a small telescope shows M33 roughly as you'd see it with binoculars, though the spiral arms are not visible without a large telescope.

One of the best-known constellations in the northern sky, Cassiopeia is always there whenever you are out stargazing.

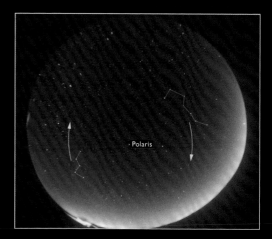

▶ Cassiopeia (left) and Ursa Major are always in the northern sky, opposite each other with the Pole Star, Polaris, in the center. They and the rest of the sky rotate around Polaris, which remains in the same place.

Where to look

Cassiopeia, with its easily spotted W-shape, is one of the constellations that lie close enough to the sky's north pole that they are always visible from most of North America. But it could be anywhere within a large area centered on the Pole Star, which is roughly halfway up the sky and due north.

In winter, Cassiopeia is pretty well overhead, so you might see it as an "M" instead of a "W," depending on which way round you stand. Then in spring it's quite high up in the northwest, but in summer it is quite low down on the northern horizon. In the fall it's climbing up in the northeast, ready for its winter appearance overhead. It is always more or less opposite the Big Dipper, with the Pole Star, Polaris, between the two of them.

Legends

In Greek myths, Cassiopeia was a vain queen, constantly preening herself and reminding everyone how lovely she was. This led to her downfall, because she really got on everyone's nerves. But Greek legends being what they are, instead of taking it out on Cassiopeia herself, the sea god punished her by chaining her daughter Andromeda to a rock where a sea monster was bound to find her. If you want to find out what happened next, you'll find the next episode in the article on **Perseus**. But eventually, Cassiopeia did get her

Point Score		
⦾⦾⦾⦾⦾⦾⦾⦾⦾⦾	date seen	points
Cassiopeia		1
NGC 663		3
Score		

See also: **NGC 457.**

Fact File

Name Cassiopeia (name of legendary character)
Area 598 square degrees

Objects	Magnitude	Distance	Type	Visibility
Shedar Alpha Cassiopeiae	2.3	228 light years	–	–
Beta Cassiopeiae	2.3	288 light years	–	–
Gamma Cassiopeiae	2.2 (variable)	612 light years	–	–
NGC 663	–	2,000 light years	Open cluster	B

come-uppance by being dangled upside down in the sky every winter, for all to see.

What you'll see

The five main stars of Cassiopeia are unmistakable, as they are all quite close together. They are all very much the same brightness except for the one on the far left of the "W," which is a bit fainter so it might be hard to spot from city centers or on a hazy night.

The star in the middle of the "W," Gamma Cassiopeiae, varies in brightness from time to time. It is a hot and unstable star, and back in the 1930s it went through swings in brightness. But since the mid 1960s it has been pretty steady. However,

▲ Here's NGC 663 as photographed through a small telescope. Use the finder chart on page 203.

it's one of those stars that astronomers keep an eye on, in case it acts up again.

There is a nice little star cluster, NGC 663, between the two left-hand stars of the "W," but a little off the line between them (see the chart opposite). You can see it with binoculars, as it is about half the size of the full Moon in the sky, but individual stars might be hard to see with small binoculars.

◄ On summer evenings Cassiopeia is low down in the northern sky.

The E.T. Cluster, NGC 457

When people see **NGC 457** through a telescope, they often laugh out loud. What makes this quite ordinary star cluster so amusing?

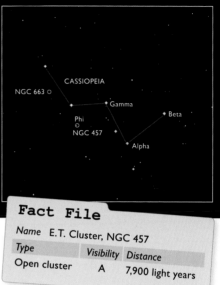

▲ To see the E.T. Cluster looking like the alien you need to see it in this orientation, with the two "eyes" at the top.

Fact File

Name	E.T. Cluster, NGC 457		
Type		Visibility	Distance
Open cluster		A	7,900 light years

Where to look

The left-hand side of the "W" of Cassiopeia is your starting point for this object. Follow the line downward and with a little bit of cheating you come to a fainter star, called Phi Cassiopeiae (often shortened to Phi Cas). This is your target, as it is the brightest star in the cluster NGC 457.

What you'll see

With binoculars, you will see very little apart from Phi Cas and another star or two. With a telescope, however, more stars appear. When the cluster is seen with south at the top, you can see a fairly bright pair of stars, one of which is Phi Cas, and several lines of stars spreading away from the two main stars. A pretty sight, which in the past was sometimes referred to as the Owl Cluster.

But think of it as the classic movie character E.T. the Extra-Terrestrial. There are the big eyes, actually glowing, and the spindly arms and legs! When you tell people that you can see E.T. through your telescope they won't believe you – until you show them NGC 457. That's when they laugh out loud!

Point Score

●●●●●●●●●●	Date seen	Points
Seeing NGC 457		2
Making someone laugh		2
Score		

A fine constellation in the northern part of the Milky Way, with plenty of interesting objects to keep you busy.

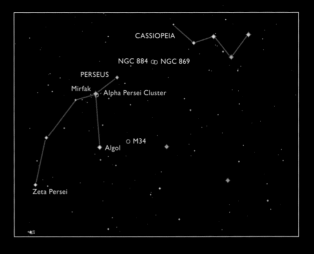

farther south. Its bright stars make a sort of lazy "T," with the bright star Mirfak at top center. The constellation extends well beyond this, with some stars stretching down toward the Pleiades in Taurus, such as Zeta Persei.

Where to look

Perseus is in the sky for much of the time, but it's particularly best seen in the evening sky in late fall and winter, when it's high in the sky. To find it, look above the signpost Square of Pegasus and spot the W-shape of Cassiopeia. Perseus is to the left of Cassiopeia and slightly

Legends

Perseus was a Greek hero, and he is shown in the sky with sword raised and the head of the Gorgon Medusa, whom he had slain, dangling from his belt. Medusa, a woman with snakes for hair, had the useful ability of turning to stone anyone who looked at her. The legend of Perseus is tied up with the neighboring constellations of Andromeda, Cassiopeia and Cetus. In one tale, he rescued the beautiful Princess Andromeda, daughter of Queen Cassiopeia, from the sea monster Cetus. He cunningly showed Medusa's head to Cetus, turning it to stone. One recent

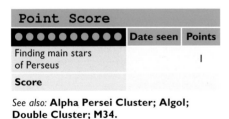

Point Score

●●●●●●●●●	Date seen	Points
Finding main stars of Perseus		I
Score		

See also: **Alpha Persei Cluster; Algol; Double Cluster; M34.**

theory suggests that the idea for this legend may have come from the fossilized bones of giant prehistoric creatures, such as early elephants, which may have been found in Ancient Greek times. Medusa must have been busy while she was alive to turn all those animals into stone!

What you'll see

Although the bright stars of Perseus are comparatively close at hand, many of the faint stars in this area are part of an outer spiral arm of the Milky Way, called in fact the Perseus Arm. It lies about 7,000 light years away. These stars all add to the general glow from the Milky Way.

Brain Box

The annual Perseid meteors, visible around August 12, appear to come from the direction of Perseus, but they have no other connection with it. They are one of the most popular meteor showers to observe.

▼ Perseus is in the Milky Way, so binoculars will reveal many groups and chains of stars that don't form actual clusters.

This is a bright cluster of stars that everyone can see but few telescopes will show!

▶ The bright star in the middle is Mirfak. Most of the stars in the S-shape and nearby are roughly 600 light years away from us and are members of the Alpha Persei Cluster.

Fact File

Name	Alpha Persei Cluster, Melotte 20	
Type	Visibility	Distance
Open cluster	A	550–650 light years

Where to look

This cluster of stars surrounds the star Mirfak, the brightest star in **Perseus**. (Use the finder chart on page 204.)

What you'll see

Mirfak is a hot, blue star about 600 light years away. With the naked eye it's just a bright star, but look at it with binoculars and you'll immediately see that it lies at the heart of a large and sparkling cluster of fainter stars, several of which are just bright enough to be seen with the naked eye, though they don't show up in typical suburban skies.

This is no coincidence – Mirfak is the brightest member of what's called the Alpha Persei Cluster. Actually, this is known technically as an association of stars rather than a cluster, because the stars are no longer very close to each other but are just traveling together through space, having formed in a cluster some 50–70 million years ago.

Look for some nice chains of stars in the cluster, in particular an S-shape that looks a bit like a rollercoaster ride. This is a cluster that can really only be seen well in binoculars – most telescopes have too high a magnification and just show you a part of the cluster at any one time. But even in light-polluted skies, the cluster is a great sight in binoculars.

Several other stars in Perseus are also moving along with the cluster, including Delta Persei and Epsilon Persei.

Point Score

⦿⦿⦿⦿⦿⦿⦿⦿⦿⦿	Date seen	Points
Finding the Alpha Persei Cluster with binoculars		2
Finding the S-shaped line of stars		2
Score		

The Double Cluster

One of the best sights in the sky, this pair of clusters is a superb sight that could have been made for viewing with small telescopes.

◀ A super sight in a dark sky, you can find the Double Cluster in binoculars even from the suburbs. But for the best results, get out into the country!

Where to look

Find **Perseus** and **Cassiopeia**, and the Double Cluster is midway between the two. If you need more help, start with the two left-hand stars of the "W" of Cassiopeia and make a long triangle from them. The Double Cluster is at the apex of the triangle. (Use the finder chart on page 204, where the Double Cluster is shown as NGC 869 and NGC 884.)

What you'll see

In good skies you can just see the Double Cluster with the naked eye. But though you can see a fuzzy patch, you can't make out individual stars. Actually, no star in either cluster is bright enough to be seen with the naked eye though the whole mass of stars together make the cluster just bright enough to be seen.

With binoculars you can see the two separate clusters, and make out stars in each one. The best views are with telescopes with low magnifications and wide fields of view, so a magnification of 20 or 30 is ideal. You may spot some red giant stars, looking slightly orange, among the fainter ones.

Point Score		
	Date seen	Points
Finding the Double Cluster with the naked eye		2
Finding the Double Cluster with binoculars		1
Spotting red giant stars		3
Score		

Fact File

Name	Double Cluster, NGC 869 and NGC 884	
Type	Visibility	Distance
Open cluster	A	7,100 and 7,400 light years

Observations of this star by a young deaf and dumb astronomer helped astronomers to understand the scale of the Universe.

▶ Make your own estimates of the brightness of Algol using this comparison star chart. Algol is marked with a circle and the constellations of Auriga, Taurus, Cassiopeia, Andromeda and Pegasus are shown in outline. Note the letters of the stars it is brighter or fainter than, then work out the brightness afterward.

A = 1.80	G = 2.89
B = 2.07	H = 3.03
C = 2.26	J = 3.17
D = 2.48	K = 3.38
E = 2.83	L = 3.54
F = 2.94	M = 3.77

Where to look

First find Mirfak in **Perseus**, then look for another star, slightly less bright, below and to its right about half a hand's breadth away. Algol is the second brightest star in Perseus – most of the time! (Use the finder chart on page 204.)

What you'll see

First, a story. It all begins in medieval times, when the civilizations of Greece and Rome had collapsed and Europe had few centers of learning. But Arab astronomers were still active, which is why so many stars today have names based on Arabic words, often beginning "Al-," meaning "The." Algol is one, and its name means "The Ghoul," perhaps because it represents the head of the mythical Medusa hanging from the belt of Perseus – the woman whose glance could turn you to stone – but perhaps because of its unusual behavior. Every so often, Algol's brightness drops by about three times, as if the Gorgon is winking at us.

The story now moves to 18th-century York and a young deaf and dumb astronomer named John Goodricke. Fortunately, John's parents were able to afford a good education for their son and didn't expect him to earn a living, so he was able to study the stars. Algol in particular fascinated him, and he frequently estimated its brightness by comparing it with other stars to try and solve the mystery of why

Point Score

⚫⚫⚫⚫⚫⚫⚫⚫⚫	Date seen	Points
Seeing Algol		1
Making a brightness estimate		3
Seeing Algol when dim		3
Score		

it was sometimes faint. He discovered that it dips in brightness regularly every 2 days 21 hours, without fail.

Goodricke decided that such a regular variation could only be caused by something orbiting Algol, such as another star that is less bright. His idea was spot on, and in fact the variations allow astronomers to measure two stars orbiting each other so close together that no telescope could show them separately. We can see the precise effects of one star covering and uncovering another, and also work out the size of the orbit and the diameters of the stars themselves.

Stars such as Algol are known as eclipsing binaries because one star of the double or binary system eclipses another. They have helped astronomers to understand very much more about stars even though they are so far away that we can see them only as points of light.

Whether you look at Algol with the naked eye, binoculars or a telescope, all you will see is an ordinary star. But to measure its brightness as John Goodricke did, you need to compare its brightness with other stars nearby which don't vary, as shown on the map. Not all the stars may be visible at the time of your observation. Decide whether Algol is brighter or fainter than each reference star in turn, until you know which pair it lies between in brightness. If you can, decide whether it is nearer one than the other, so you can improve your estimate. For help on the magnitude scale, see page 16.

Repeat your observation from time to time to get further estimates – but to do what Goodricke did, you'd have to do the same thing very frequently until you found enough dips in Algol's brightness to be sure of what it was doing.

▼ Algol varies because a dimmer star orbits a brighter one. This shows how its orbit matches its magnitude variations hour by hour.

Fact File

Name		Magnitude	Distance
Algol	Beta Persei	2.1–3.4 (variable)	90 light years

The star cluster M34 is one of the brightest in autumn and winter skies, and it is quite easy to find with binoculars even if there's light pollution.

▶ The star cluster M34 in Perseus. See if you can spot it with the naked eye once you've found it with binoculars.

Fact File

Name	M34		
Type		Visibility	Distance
Open cluster		A	1,500 light years

Where to look

If you can find **Algol**, you can find M34, because it's just a short distance from that star. Though M34 can be seen with the naked eye under good conditions, most people will need binoculars to find it. So first find Algol, and also look for the nearest star to its west, which is called Gamma Andromedae or Almach. This is about the same brightness as Algol unless Algol is in one of its fades (see page 208).

Now scan with binoculars between Algol and Almach, and you should pick up M34 roughly midway between the two of them. You may find that you can see both Algol and M34 in the same field of view. (Use the finder chart on page 204.)

What you'll see

None of the stars in M34 are particularly bright, and none can be seen individually with the naked eye. The brightest are about

Point Score

●●●●●●●●●●	Date seen	Points
Finding M34		1
Counting the stars		3
Score		

ten times fainter than you can see without help even under good conditions, so you might wonder how M34 can ever be seen with the naked eye. But the combined light of many stars all adds up, so in really good skies you can see just a little hazy area.

Altogether, M34 has several hundred stars, but only about 20 are so are usually visible with binoculars. Try to count them! In general most of the ones you see will be in M34 itself, though some will be closer or more distant and in the same line of sight.

Cepheus
Mu Cephei and
Delta Cephei

99

**See if you can repeat the
observations of a star that
made a teenager famous.**

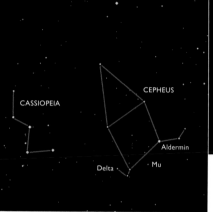

CASSIOPEIA

CEPHEUS

Aldermin

Delta · Mu

Where to look

Like Cassiopeia, Cepheus is always visible
from North America because it is quite
close to the sky's north pole. If you are
looking at Cassiopeia so that it is a "W,"
Cepheus is to its right. The right-hand side
of the "W" points straight at Alderamin,
Alpha Cephei, about three side-lengths
away. It's at its highest in the fall, when it
is virtually overhead from northern areas.

The main stars of Cepheus form a sort
of wonky house-shape, with a rectangle for
the house and a long triangle for the roof,
and a couple of little triangles at the sides
along the bottom.

Legends

King Cepheus was a mythical king married
to the vain Queen Cassiopeia, who is of
course right next to him in the sky. The
story so far is that their daughter, the
beautiful Andromeda, was chained to a
rock by the sea and was about to be eaten
by the sea monster Cetus when Perseus
saved her. Now read on….

So Perseus claimed Andromeda as his
bride and a wedding was arranged. The
only fly in the ointment was Cepheus'
brother, Phineus, who was under the belief
that he was going to marry Andromeda.
In the best traditions of soap operas, there
was an almighty bust-up at the wedding.
Perseus was seriously outnumbered, but

Point Score

●●●●●●●●●	Date seen	Points
Finding Cepheus		1
Finding Mu Cephei		1
Finding Delta Cephei		1
Measuring changes in Delta's brightness		4
Score		

of course he had the head of Medusa with him, which turned all his assailants, including the luckless Phineus, to stone. The story ends with King Cepheus and Queen Cassiopeia being placed in the heavens. Cepheus' poor judgment at letting the whole thing get out of hand is rewarded by being given forever a dunce's cap, which is the long triangle that makes the roof of the house.

What you'll see

There are plenty of stars in Cepheus, but no bright deep-sky objects. Two stars in particular are worth a look. One is Mu Cephei, which is in the garden of the house, just below the rectangle. This is described by astronomers as a

▼ Mu Cephei, the Garnet Star. Make up your own mind how red it really is!

very red star, so you might expect to see something really striking, but what astronomers call red, anyone else would call white with just a bit of a blush about it.

The best way to see the color is to view it through binoculars or a telescope. The great William Herschel, the 18th-century astronomer who discovered the planet Uranus and many deep-sky objects, said that it was garnet in color, so it's often referred to as the "Garnet Star." He suggested looking first at Alpha Cephei then Mu to get the best effect.

Fact File

Name	Cepheus (name of legendary character)		
Area	588 square degrees		
Objects		Magnitude	Distance
Alderamin	*Alpha Cephei*	2.5	49 light years
Garnet Star	*Mu Cephei*	4.0	6,000 light years
	Delta Cephei	3.5–4.4	860 light years

Like many other red stars, this is a red supergiant, and like all such stars is very bloated. Exactly how big isn't certain, because its distance is not known very well, but it is probably as big as the orbit of Jupiter around the Sun. And also like other red supergiants, one day it will probably explode as a supernova, like the **Crab Nebula**, and become a brilliant object in our skies.

The other interesting star in Cepheus is Delta Cephei, which is in the triangle at the left-hand base of the house. Its brightness varies from magnitude 3.5 to 4.4 every 5.3 days. This isn't such a large change that you would notice it straight away, but it was measured by John Goodricke back in the 18th century, following his great success while still a teenager in discovering that Algol varied on a regular basis.

Though Goodricke couldn't know it at the time, Delta Cephei and stars like it (called Cepheids, after Delta Cephei) were to play a vital role in working out the size of the Universe. After many other such stars were found, in the 20th century it turned out that the slower their variations, the brighter they are. So as soon as you notice that a star varies in the same regular way as Delta Cephei, all you need to do is see how quickly or slowly it varies and you know how bright it really is. Compare that with the brightness it appears in the sky, and you can work out its distance.

One of the reasons why the Hubble Space Telescope was built was to find Cepheids in the galaxies of the Virgo Cluster, which are just at the limit of its cameras. This meant that astronomers could measure the distances of these galaxies quite accurately, which in turn helped to work out the distances to other galaxies.

▲ Delta Cephei is an attractive wide double star as well as being a variable. You should see its sixth-magnitude companion star with binoculars.

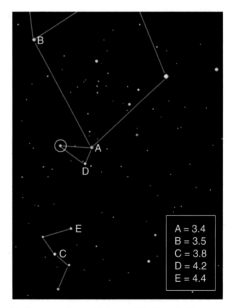

| A = 3.4 |
| B = 3.5 |
| C = 3.8 |
| D = 4.2 |
| E = 4.4 |

▲ Compare the brightness of Delta with the stars used by Goodricke, to see if you can follow the variations of Delta Cephei (shown by a circle).

213

100

Ursa Minor, the Little Bear
Polaris

Once you've found Polaris and Ursa Minor you shouldn't lose them as they always stay in the same place!

Where to look

As the brightest star in Ursa Minor, Polaris is your best guide to finding the constellation. It is less than a degree from the sky's north pole, which means that it always stays in almost exactly the same spot. So look due north, and at an angle equal to your latitude, which will be somewhere between 25° and 60° above the horizon, and there it is. It's often called the North Star or Pole Star, and people often think that because it's well known it should be particularly bright, but in fact it's only second magnitude.

Point Score		
⊙ ⊙ ⊙ ⊙ ⊙ ⊙ ⊙ ⊙ ⊙	**Date seen**	**Points**
Finding Ursa Minor		1
Seeing the "Engagement Ring" around Polaris		2
Score		

The stars of Ursa Minor trail away from Polaris and end in a rectangle, so they look a little like the Big Dipper, and the group is often referred to as the Little Dipper.

The star Kochab at the other end of the line from Polaris is almost the same brightness, so there's a risk of confusing the two, particularly in spring and late summer when the two stars are roughly the same height above the horizon. But there's another way to find Polaris: use the

Fact File

Name	Ursa Minor, the Little Bear		
Area	256 square degrees		
Objects			
		Magnitude	Distance
Polaris	Alpha Ursae Majoris	1.79	432 light years

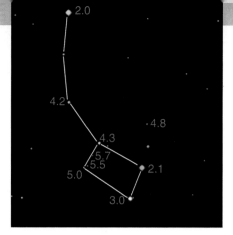

▲ The magnitudes of some of the stars in Ursa Minor. Use it to judge how dark your sky is.

two right-hand stars of the Big Dipper, Merak and Dubhe. Follow the line of these two stars and they point (well, nearly) to Polaris. This has earned them the name of "the Pointers."

Legends

There are few stories about Ursa Minor, and in fact the Greeks used to think of it as a dog. But some tales link it to the story of Callisto and Arcas, who became Ursa Major and Boötes, and say that Ursa Minor is actually Arcas. Because Ursa Minor has seven stars in roughly the same pattern as Ursa Major, it makes sense to think of them both as bears, even if they have unfeasibly long tails.

What you'll see

Polaris is best known for its position very close to the pole, which makes it one of the most useful stars there is, and one which every astronomer (in the northern hemisphere at least) ought to know. There is no similar star close to the sky's south pole, so we can feel rather smug (though people in the southern hemisphere have a lot to be smug about, as there are some amazing sights which we never get to see).

Look at Polaris with binoculars on a fairly clear night and you'll notice that there is a sixth-magnitude star close to it, and a ring of fainter stars extending farther away. These stars are not connected in any way, but together they are romantically referred to as the "Engagement Ring."

One advantage of the stars in this part of the sky is that because they are always in virtually the same position, they are useful for measuring the faintest star that you can see – what's known as the *limiting magnitude*. Of course there are plenty of other stars in the sky that you could use for this, such as those in the Square of Pegasus, but they aren't always around or they may be low in the sky so you won't get consistent results.

In Ursa Minor there are examples of stars of second, third, fourth and fifth magnitude. Use the map here to decide which is the faintest star you can see. If you do the same every time you observe, you can quickly discover which are the best nights for observing. As always, wait until you are properly dark adapted before you make your measurement.

▲ A close-up of Polaris (in the middle) and the "Engagement Ring" of stars nearby. This view is about 3° across – about the same as high-power binoculars.

215

101

The Milky Way ◎🔭

These days, many people never get to see the Milky Way because of light pollution. But get away from the cities and it's still there!

▶ The Milky Way can be seen well from many country areas – you don't have to go to the remotest parts of the world.

Point Score

●●●●●●●●●●	Date seen	Points
Seeing the Milky Way		3
Seeing it really well		3
Score		

Where to look

Although the Milky Way stretches all the way around the heavens and is always in the sky somewhere, the best time to look for it is in late summer into the fall. At this time of year it is high up, and the brightest parts of it are on show from overhead down to the southern horizon. During winter and spring it is also overhead in the early evening, but this part of it is less bright and you need darker skies to see it.

In late spring and early summer it is hard to see as it runs along the horizon from the west through north to east, so even from country sites it is hidden by the light pollution from towns some distance away.

What you'll see

To the eye, the Milky Way is a broad misty band stretching across the sky. Actually, it looks a bit like a band of pale cloud, as it has an irregular outline, and people are quite often fooled into thinking that the clouds are coming over when in fact it's their first view of the Milky Way under really dark conditions!

In fact you don't need perfect conditions to be able to see it at the right time of year. It's visible from many country locations, and the key thing is to get away from the really big cities, which cast their glow for 80 kilometers or more. But there's a big difference between just being able to see it and getting a real eyeful, under good conditions. Then it seems so bright that you wonder how you could ever miss it normally.

You'll notice that the Milky Way is not an even brightness. There are brighter and fainter areas, and bits where there seems to be some sort of obscuration. The most famous, from the northern hemisphere at least, is the **Great Rift**, which starts in Cygnus and finishes up in Ophiuchus and Sagittarius. Down in the southern hemisphere there is a smaller but darker area known as the Coal Sack. These are caused by dark clouds of dust and gas in space.

With binoculars, you can see that the Milky Way is actually the effect of millions of faint stars. It's fun just to sweep along the Milky Way and see the groups and clusters. Some of these you might recognize because they are in this book, but there are many other smaller groupings that don't usually get a mention because they have fewer stars in them. Through a telescope, the effect of all the stars gets rather lost, because its field of view is usually smaller than that of binoculars.

The Milky Way is really our own galaxy seen from inside. The Sun is fairly centrally placed within its thickness, but a long way out from the center. This is why it is brighter in the direction we see in summer, through Sagittarius, than in winter, when it passes through Orion.

The band of the Milky Way contains most of the gas and dust clouds, the star clusters and the other objects, such as planetary nebulae that are the result of stars reaching the ends of their lives. The dust clouds hide a lot of it. We can't see stars that are on the far side of the Milky Way, because there is just too much material in the way. Most of the individual stars that you can see with the naked eye are fairly close to us, within a few hundred light years.

▼ An artwork of what the Milky Way might look like if we could photograph it from the outside. There is a central bar with two main spiral arms and several smaller ones. The Sun is somewhere in the small circle, in a minor spiral arm known as the Orion Spur, because it also contains the Orion Nebula.

Fact File

Name	Milky Way
Visibility	C
Diameter	120,000 light years
Distance of Sun from center	28,000 light years
Number of stars	200–400 billion stars

STAR MAPS

The star maps on the next few pages show the whole of the sky that you will see from North America. But you'll only see part of it at a time, and to find out which part you need to choose the right maps.

Maps on these pages

The circular map here covers the sky looking north. In this direction, the sky turns counterclockwise around the center of the map, very close to the star Polaris. It takes slightly less than 24 hours to do so, which means that you'll see it a different way round from month to month.

Hold the map with the current month at the bottom, and it shows you the stars in the correct orientation for that month at 9 pm in winter time or 10 pm during Daylight Saving Time. But if you want to see the sky **earlier**, just turn it a month earlier for every two hours earlier. So if you want to observe at 7 pm in December, for example, hold the map with November at the bottom.

The height of Polaris above your horizon depends on your latitude. The farther south you live, the lower it will appear in the sky. From Anchorage, Alaska, it is at 60° above the horizon, which is on the bottom of the map, but from Florida it is at 25° up and you won't see some stars at the bottom of the map.

Sky looking north

218

What's on the maps

Each object covered in this book is shown, other than the Solar System bodies which are always on the move and can't be featured. The different types of object – star clusters, planetary nebulae, bright nebulae or galaxies – are shown by different symbols according to the key above. However, the objects are not usually as large as the symbols that depict them so refer to the pages that describe the objects to find out what you should be looking for.

Stars that vary in brightness are shown by circles that indicate the brightest and faintest magnitude that the star can be. So a star that is sometimes bright, but sometimes too dim to be shown on the charts, is shown as an open circle.

Also marked are the actual constellation boundaries. The Greek symbols by the stars show their Bayer letters (see page 8). The fainter stars are given numbers, known as Flamsteed numbers after the English astronomer who drew up a catalog published in 1712. Mostly these are in the correct constellation, but see if you can find an example of a star which Flamsteed numbered within a different constellation. Hint: sometimes constellation names are abbreviated to three letters, such as UMa for Ursa Major.

The pale blue lines are the sky's grid system of Right Ascension and Declination (see page 19).

219

Spring stars

Maps on these pages

On these next four pages are maps that show the stars looking south. But in this case, you don't need to turn the book around, as the stars appear the right way up all the time. Again, the sky moves over the course of the year, and you might think that the maps are shown in the wrong order, with winter following spring. But this is because the stars move from east to west over the course of a year as well as from hour to hour.

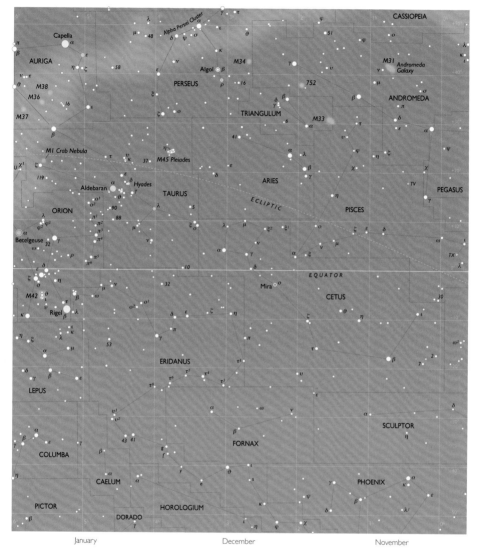

So choose your month and look south. The stars directly above the name of the month will be those that you see looking south at 9 pm in winter or 10 pm in summer. The top of the seasonal map overlaps with the top of the northern skies map.

If you want to observe earlier in the evening, look one month earlier for each two hours earlier you are observing. The stars move quite slowly, so it should be easy enough to get your bearings. And if you get up in the middle of the night,

Autumn stars

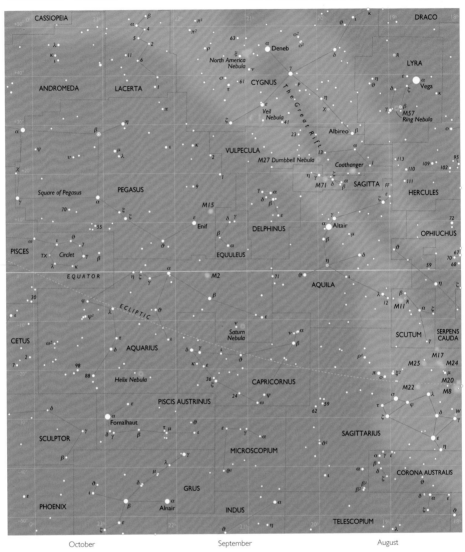

October   September   August

you can see stars which won't be in the evening sky for several months.

The stars to one side or other of the month you've chosen will be at a bit of an angle to the horizontal, but you should be able to pick them out all the same.

The position of the bottom of the map will depend on your latitude. Some stars are shown which are always below the horizon from most of the United States. So look for the stars in the middle of the map rather than

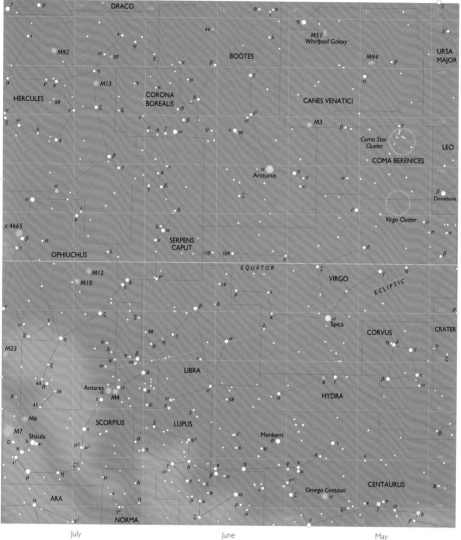

Star clusters   Planetary nebulae   Bright nebulae   Galaxies

July   June   May

at the bottom. Bear in mind that the top of the maps will be almost overhead.

In addition to the 34 constellations that are described in detail in this book because there are interesting objects there, you will find many others. There are 88 constellations in all, including 12 in the part of the sky that isn't shown on these maps because it never rises from most of North America.

INDEX